STUDENT SOLUTIONS MANUAL FOR
CALCULUS
from Graphical, Numerical, and Symbolic Points of View

VOLUME 2

ARNOLD OSTEBEE AND PAUL ZORN
St. Olaf College

1995
PRELIMINARY
EDITION

SAUNDERS COLLEGE PUBLISHING
Harcourt Brace College Publishers

Fort Worth Philadelphia San Diego New York Orlando Austin
San Antonio Toronto Montreal London Sydney Tokyo

This material is based upon work supported by the National Science Foundation under Grant No. USE-9053363.

Any opinions, findings, and conclusions or recommendations expressed in this material are those of the author and so do not neccessarily reflect the views of the National Science Foundation.

Ostebee & Zorn; Student Solutions Manual for <u>Calculus From Graphical, Numerical, and Symbolic Points of View, Vol. II, 1995 Preliminary Edition.</u>

ISBN 0-03-017589-5

567 017 987654321

page 5, first margin note: SHOULD read ``requiring that f and g
         be continuous assures ... ''
         ^^^^^^^^^^^^^

page 8, line 1: SHOULD read ``... s_A and s_b '', not
         ``... s_At and s_B ... ''

page 54, picture of Region 1: the shading SHOULD be only
         to the RIGHT of the y-axis, i.e., from x=0 to x=3.

Page 81, Exercise 11: Integrand should have one dx, not two.

Page 95, line -3: the second mention of K_1 in Example 3
         is missing the subscript. It SHOULD read ``K_1'', not ``K''.

Page 119, second paragraph, third line: SHOULD read
         ``... at constant speed ...''
                      ^^^^^

Page 121, second margin note, first line: ``Antiderivatives''
         should be plural.

Page 124, exercise 4d: Integral formula needed is #38, not #34.

Page 130, exercise 18: SHOULD read ``uniform circular cross-sections'',
         not ``circular cross-sections''.

Page 130, exercise 19: SHOULD read ``circumference'', not ``radius''.

Page 137, exercises 1 and 2: SHOULD read ``to x=1 using the FACT on
         p. 136''

Page 143, line 4: change ``90'' to ``30''.

Page 163, exercises 29 and 30: add ``dx'' to both integrals

Page 178, third picture of a triangle: interchange labels on
         horizontal and vertical sides.

Page 186, exercise 1(f): the numerator of the integrand SHOULD
         be ``cos x'' not ``sin x''

Page 190, line -2: SHOULD be M_50, not M_20.

Page 196 - 7, exercises 6b, 7b, 8, 18, 19, 20: Integrands should
         have one ``dx'', not two.

Page 197, exercise 8(b) and (d): The inequalities ``x >= 1''
         SHOULD read ``x >= e''.

Page 214, line -4: SHOULD read ``of Example 1'', not
         ``of Example harmseqex''.

Page 216, in the Fact: DELETE one ``then''.

Page 228, line -7: SHOULD read ``r > 1'', not ``r >= 1''.

Page 232, exercise 3: summation should start at k=1, not k=0.

Page 235, exercise 28: change all k's to m's.

Page 235, exercise 31: change all k's to j's.

Page 238, exercise 55(h): SHOULD refer to parts ``(f) and (g)''
    not ``(d)'' and ``(e)''

Page 249, exercise 37: summation should start at ``k=1'', not ``k=0''

Page 256, lines 2 and 3: The second sentence SHOULD read: ``The
    given series IS alternating, but another hypothesis isn't
    satisfied ... '' .

Page 258, exercise 24 (a): All n's SHOULD be k's.

Page 262, line -2: SHOULD be ``1+2x+3x^2 ...'', not ``1+2x+3x^3 ''.
                                       ^^^   ^^^
Page 275, exercise 9: SHOULD read ``f(x) = ... '', not ``(x)= ...''.

Page 276, exercise 37: SHOULD read ``x -> 1'' not ``x -> 0''

Page 277, exercise 44: SHOULD read ``1 - cos x < ln (1+x) < ...''
    not ``1 - cos x ln (1 + x) ...''

Page 310, third line of Example 2: CHANGE ``P - 10000'' to
    ``10000 - P''.

Page 318, line -5:   first denominator needs a right parenthesis

Pages 324, exercise 9, line 2:: CHANGE ``P - 10000'' to
    ``10000 - P''.

Page 332, bottom: Coordinates of center are (3/2,0), not (0,3/2).

5.1   Areas and integrals

1. The area of the entire rectangle shown is only 250, so $\int_1^2 g(x)\,dx < 250$ must be true. Furthermore, since $g(x) > 0$ for all x, $\int_1^2 g(x)\,dx > 0$ must be true. Finally, it is clear from the picture that $\int_{1.75}^2 g(x)\,dx > 12.5$ (the area of one dotted rectangle). Thus, the only possible choice is $\int_1^2 g(x)\,dx \approx 45$.

3. Jack's answer is too big — the area of the entire rectangle shown is only $\pi/2$. Since $\cos^8 x \geq 0$ on the entire interval $[0, \pi/2]$, the value of the integral must be positive. This rules out Ed's answer. Finally, the value of the integral is approximately the area of the triangle with vertices at $(0, 0)$, $(0, 1)$, and $(1, 0)$. Since this triangle has area $1/2$, Lesley's answer is too big. Therefore, Joan's answer must be the correct answer.

    [NOTE: Since $\cos^8 x \leq 1$ when $0 \leq x \leq 1/2$ and $\cos^8 x \leq 2/5$ when $1/2 \leq x \leq 1$, $\int_0^1 \cos^8 x\,dx = \int_0^{1/2} \cos^8 x\,dx + \int_{1/2}^1 \cos^8 x\,dx \leq 1/2 + 1/5 = 7/10$.]

5. (a) $h'(5) \approx -2.2$ (d) $\int_0^{10} h(x)\,dx \approx 17$

    (b) $\frac{1}{10}\int_0^{10} h(x)\,dx \approx 17/10 = 1.7$ (e) $\int_0^5 h(x)\,dx \approx 20$

    (c) $\frac{1}{10}\big(h(10) - h(0)\big) = -4/10 = -0.4$ (f) $\int_6^{10} h(x)\,dx \approx -4.7$

7. (a) $\int_0^4 v(t)\,dt = 80$. At time $t = 4$ the car is 80 miles east of its starting point.

    (b) The car's average (eastward) velocity is 20 mph $= \frac{1}{4}\int_0^4 v(t)\,dt$.

    (c) Since $s(t) = |v(t)|$, $\int_0^4 s(t)\,dt = 100$. Over the 4 hours, the car travels a total distance of 100 miles.

    (d) 100 miles/4 hours $= 25$ mph

9. (a) Since $2 < f(x) < 5$ on $[1, 6]$, $10 < \int_1^6 f(x)\,dx < 25$ must be true. Thus, the best estimate for the value of the integral is 20. (NOTE: Be sure to take into account that the interval of integration is $[1, 6]$ and not $[0, 8]$!)

    (b) $A = 12$ and $B = 20$ since $3 \leq f(x) \leq 5$ when $3 \leq x \leq 7$. (Many other answers are possible.)

    (c) This approximation underestimates the exact value of the integral. (Count squares.)

    (d) $\frac{1}{2}\int_0^2 f(x)\,dx \approx \dfrac{11}{2}$

11. (a) $\int_0^2 f(x)\,dx = 6$, $\int_1^4 f(x)\,dx = 45/2$, $\int_{-5}^{-1} f(x)\,dx = -36$, $\int_{-2}^3 f(x)\,dx = 15/2$

    (b) $\int_0^2 f(x)\,dx = 14$, $\int_1^4 f(x)\,dx = 30$, $\int_{-5}^{-1} f(x)\,dx = -4$, $\int_{-2}^3 f(x)\,dx = 30$

    (c) $\int_0^2 f(x)\,dx = 6$, $\int_1^4 f(x)\,dx = 0$, $\int_{-5}^{-1} f(x)\,dx = 44$, $\int_{-2}^3 f(x)\,dx = 20$

13. (a) The trapezoid with vertices $(2, 0)$, $(2, 3)$, $(4, 2)$, and $(4, 0)$ has area 5 and lies below the graph of f over the interval $[2, 4]$.

    (b) The trapezoid with verticles $(6, 0)$, $(6, -2)$, $(9, -4)$, and $(9, 0)$ has area 9.

    (c) Yes, since $\int_5^6 f(x)\,dx < -1.25$ (draw a triangle with vertices at $(5, 0)$, $(6, 0)$, and $(6, -2.5)$) and $\int_0^5 f(x)\,dx = 4 + 9\pi/4 \approx 11.069 < 11.25$.

    (d) $\int_2^7 f(x)\,dx > 0$ because $\int_2^5 f(x)\,dx = 9\pi/4 > 7$ and $\int_5^7 f(x)\,dx > -6$ (draw a trapezoid with vertices at $(5, 0)$, $(5, -2)$, $(7, -4)$, and $(7, 0)$).

15. $\int_0^2 f(x)\,dx = \int_0^1 f(x)\,dx + \int_1^2 f(x)\,dx = \frac{3}{2} + \frac{1}{2} = 2$

17. $\int_1^3 \left(6 - \sqrt{4 - (x - 3)^2}\,dx\right) dx = 12 - \pi$ [NOTE: $y = -\sqrt{4 - (x - 3)^2}$ is the bottom half of a circle with radius 2 and center at $(3, 6)$.]

19. The regions whose areas are represented by the three integrals are horizontal translates of each other.

21. (a) $\int_1^4 f(x)\,dx = \int_1^2 f(x)\,dx + \int_2^4 f(x)\,dx = -1 + 7 = 6$

    (b) $\int_0^4 3f(x)\,dx = 3\left(\int_0^2 f(x)\,dx + \int_2^4 f(x)\,dx\right) = 3 \cdot 9 = 27.$

    (c) $\int_0^1 f(x)\,dx = \int_0^2 f(x)\,dx - \int_1^2 f(x)\,dx = 2 - (-1) = 3.$

    (d) $\int_0^1 f(x+1)\,dx = \int_1^2 f(x)\,dx = -1$

    (e) $\int_0^1 \left(f(x) + 1\right)\,dx = \int_0^1 f(x)\,dx + \int_0^1 dx = 3 + 1 = 4.$

    (f) $\int_2^4 f(x-2)\,dx = \int_0^2 f(x)\,dx = 2.$

    (g) $\int_2^4 \left(f(x) - 2\right)\,dx = \int_2^4 f(x)\,dx - 2\int_2^4 dx = 7 - 2 \cdot 2 = 3.$

    (h) $-1 = \int_1^2 f(x)\,dx < 0$ implies that $f(x) < 0$ over some (or all) of the interval $[1, 2]$.

    (i) $6 = \int_2^4 3\,dx < \int_2^4 f(x)\,dx = 7$ implies that $f(x) > 3$ over some (or all) of the interval $[0, 2]$.

    (j) One possibility: $f(x) = \begin{cases} -8(x-1) - 1, & 0 \le x \le 1 \\ -1, & 1 < x \le 2 \\ 9(x-2) - 1, & x > 2 \end{cases}$

23. (a) $\int_0^\pi \cos x\,dx = 0$ (c) $\int_{-2}^2 \left(7x^5 + 3\right)\,dx = 12$

    (b) $\int_{\pi/2}^{3\pi/2} \sin x\,dx = 0$ (d) $\int_{-1}^1 \left(4x^3 - 2x\right)\,dx = 0$

25. $\displaystyle\int_1^3 \frac{1-x}{x^2}\,dx = \int_1^2 \frac{1-x}{x^2}\,dx + \int_2^3 \frac{1-x}{x^2}\,dx.$ Since the integrand $((1 - x)/x^2)$ is negative over the interval $[2, 3]$

$\displaystyle\int_2^3 \frac{1-x}{x^2}\,dx < 0.$

27. $\displaystyle\int_a^b f(x)\,dx \ge \int_a^b m\,dx = m(b - a)$ and $\displaystyle\int_a^b f(x)\,dx \le \int_a^b M\,dx = M(b - a).$

29. $0 \le x \sin x \le x$ when $0 \le x \le \pi$, so $0 \le \int_0^\pi x \sin x\,dx \le \int_0^\pi x\,dx = \pi^2/2$

31. Since $-1 \le \cos x \le -1/2$ when $2\pi/3 \le x \le \pi$, $-(\pi - 2\pi/3) \le \displaystyle\int_{2\pi/3}^\pi \cos x\,dx \le -\frac{1}{2}(\pi - 2\pi/3) = -\pi/6$

33. (a) Let $f(x) = e^x - (1 + x)$ and $g(x) = 1 + 3x - e^x$. Then $f'(x) = e^x - 1 \ge 0$ and $g'(x) = 3 - 3^x \ge 0$ when $0 \le x \le 1$. Since $f(0) = g(0) = 0$, the inequalities $1 + x \le e^x \le 1 + 3x$ are valid when $0 \le x \le 1$.

    (b) Since $\int_0^1 x\,dx = 1/2$, the inequalities in part (a) imply that $3/2 \le \int_0^1 e^x\,dx \le 5/2.$

35. (a) Since $1/2 \le \cos x \le 1$ when $0 \le x \le \pi/3$, $\pi/6 \le \displaystyle\int_0^{\pi/3} \cos x\,dx \le \pi/3.$ Similarly, since $0 \le \cos x \le 1$ when

    $\pi/3 \le x \le \pi/2, 0 \le \displaystyle\int_{\pi/3}^{\pi/2} \cos x\,dx \le \pi/12.$ Thus, $\pi/6 \le \displaystyle\int_0^{\pi/2} \cos x\,dx \le \pi/3 + \pi/12 = 5\pi/12.$

    (b) $\displaystyle\int_0^{\sqrt{\pi/2}} \cos x^2\,dx = \int_0^{\sqrt{\pi/3}} \cos x^2\,dx + \int_{\sqrt{\pi/3}}^{\sqrt{\pi/2}} \cos x^2\,dx$ so

$$\frac{1}{2}\sqrt{\frac{\pi}{3}} \le \int_0^{\sqrt{\pi/2}} \cos x^2\,dx \le \sqrt{\frac{\pi}{3}} + \frac{1}{2}\left(\sqrt{\frac{\pi}{2}} - \sqrt{\frac{\pi}{3}}\right) = \frac{1}{2}\left(\sqrt{\frac{\pi}{2}} + \sqrt{\frac{\pi}{3}}\right)$$

37. $\displaystyle\int_0^{\pi/2} \cos x\,dx > \int_0^{\pi/4} \cos(2x)\,dx.$ Since $\int_{\pi/4}^{\pi/2} \cos x\,dx > 0$ and $\cos(2x) \le \cos x$ when $0 \le x \le \pi/4$,

$\displaystyle\int_0^{\pi/2} \cos x\,dx = \int_0^{\pi/4} \cos x\,dx + \int_{\pi/4}^{\pi/2} \cos x\,dx > \int_0^{\pi/4} \cos x\,dx \ge \int_0^{\pi/4} \cos(2x)\,dx.$

39. $\sqrt{1 + \cos(2x)} = \sqrt{2\cos^2 x} = \sqrt{2}\cos x$ when $0 \le x \le \pi/2$.

41. Since f is an odd function, $\int_{-a}^{0} f(x)\,dx = -\int_{0}^{a} f(x)\,dx$. Thus, $\int_{-a}^{a} f(x)\,dx = 0$.

43. The bounds on f imply that $-4 \le \int_{1}^{3} f(x)\,dx \le 10$. Thus, $-2 \le \dfrac{\int_{1}^{3} f(x)\,dx}{2} \le 5$.

45. $\int_{-3}^{1} f(x)\,dx = 2 \cdot 4 = 8$ and $\int_{-3}^{7} f(x)\,dx = 10 \cdot 5 = 50$, so $\int_{1}^{7} f(x)\,dx = 42$. Therefore, the average value of f over the interval $[1, 7]$ is $42/6 = 7$.

47. Both integrals measure the area of the same region.

49. (a) No. Let $f(x) = 0$ and $g(x) = x$. Then $\int_{-1}^{1} f(x)\,dx = \int_{-1}^{1} g(x)\,dx = 0$ but $f(x) \ge g(x)$ when $-1 \le x \le 0$.

    (b) Yes. If $f(x) > g(x)$ for every x such that $a \le x \le b$, then $\int_{a}^{b} f(x)\,dx > \int_{a}^{b} g(x)\,dx$ (i.e., a contradiction).

51.

53.

5.2   The area function

1. Suppose that $x < 0$. Then, $\int_x^0 t\,dt = -x^2/2$ and, therefore $\int_0^x t\,dt = x^2/2$.

3. (a) $F(x) = ax$; $G(x) = a(x-2)$; $H(x) = a(x+1)$. Yes.
    (b) $F(x) = bx^2/2$; $G(x) = b(x^2-4)/2$; $H(x) = b(x^2-1)/2$. Yes.
    (c) $F(x) = bx^2/2 + ax$; $G(x) = b(x^2-4)/2 + a(x-2)$; $H(x) = b(x^2-1)/2 + a(x+1)$. Yes.

5. (a) $A_f(\pi) = 2$, $A_f(3\pi/2) = 1$, $A_f(2\pi) = 0$, $A_f(-\pi/2) = 1$, $A_f(-\pi) = 2$, $A_f(-3\pi/2) = 1$, $A_f(-2\pi) = 0$
    (b) Since f is 2π-periodic, $A_f(x) = \int_0^x f(t)\,dt = \int_{2\pi}^{x+2\pi} f(t)\,dt$. Now, $A_f(2\pi) = 0$, so $\int_{2\pi}^{x+2\pi} f(x)\,dx = \int_0^{x+2\pi} f(x)\,dx$ $A_f(x+2\pi)$. Thus, $A_f(x) = A_f(x+2\pi)$ which implies that A_f is 2π-periodic.
    (c) Since f is positive on the interval $[0, \pi]$, A_f is an increasing function on this interval. Thus, $A_f(0) = 0 \le A_f(x) \le A_f(\pi) = 2$ when $0 \le x \le \pi$. Since f is negative on the interval $[\pi, 2\pi]$, A_f is decreasing on this interval. Thus, $A_f(\pi) = 2 \ge A_f(x) \ge A_f(2\pi) = 0$ when $0 \le x \le \pi$. Finally, since A_f is 2π-periodic, we may conclude that $0 \le A_f(x) \le 2$ for all x.
    (d) $A_f(x) = 1 - \cos x$

7. (a) $\displaystyle\int_{\sqrt{\pi/2}}^x f(t)\,dt = \int_0^x f(t)\,dt - \int_0^{\sqrt{\pi/2}} f(t)\,dt = \sin x^2 - \sin(\pi/2) = \sin x^2 - 1$
    (b) $\displaystyle\int_{-\sqrt{3\pi/2}}^x f(t)\,dt = -\int_0^{-\sqrt{3\pi/2}} f(t)\,dt + \int_0^x f(t)\,dt = -\sin(3\pi/2) + \sin x^2 = 1 + \sin x^2$

9. (a) f is non-negative on $[a, b]$
    (b) F is decreasing on $[a, b]$
    (c) F is concave up on $[a, b]$
    (d) $f'(x) \le 0$ when $a \le x \le b$
    (e) $G(x) = F(x) + C$ where C is a constant

11. (a) $A_f(5)$ is larger because f is positive on the interval $[1, 5]$
    (b) $A_f(7)$ is larger because f is negative on the interval $[7, 10]$
    (c) $A_f(-2) < A_f(-1) < 0$
    (d) A_f is increasing on the interval $(-2, 6)$
    (e) It is a local maximum because f changes from positive to negative (i.e., A_f changes from increasing to decreasing).
    (f) $A_f(x) = \displaystyle\int_0^x f(t)\,dt = \int_{-2}^x f(t)\,dt - \int_{-2}^0 f(t)\,dt = F(x) + C$ where $C = -\int_{-2}^0 f(t)\,dt$. $C < 0$ because $f(x) >$
        when $-2 \le x \le 0$.
    (g) $0 < A_f(0) - A_f(-1) < A_f(-1) - A_f(-2) < A_f(1) - A_f(0) < A_f(2) - A_f(1)$
    (h) These values suggest that A_f is concave down on the interval $[3, 8]$—the slopes of the secant lines are decreasing.

13. (a) Suppose that $a < x < b$. Then $\dfrac{A_f(x) - A_f(a)}{x - a} = \dfrac{\int_a^x f(t)\,dt}{x - a} < f(x)$ since f is increasing on the interval $[a, x]$
        Similarly, $\dfrac{A_f(b) - A_f(x)}{b - x} = \dfrac{\int_x^b f(t)\,dt}{b - x} > f(x)$ since f is increasing on the interval $[x, b]$. Thus, for any x such that
        $a < x < b$, $\dfrac{A_f(x) - A_f(a)}{x - a} < \dfrac{A_f(b) - A_f(x)}{b - x}$ which implies that A_f is concave up on the interval $[a, b]$.
    (c) Since the concavity of A_f changes at a, A_f has an inflection point at a.

15. $G(x) - F(x) = \displaystyle\int_b^x g(t)\,dt - \int_a^x f(t)\,dt = \int_b^c g(t)\,dt + \int_c^x g(t)\,dt - \int_a^c f(t)\,dt - \int_c^x f(t)\,dt$
    $= G(c) - F(c) + \displaystyle\int_c^x \big(g(t) - f(t)\big)\,dt \ge 0$ when $x \ge c$.

17. (a) $A_f(x) = \displaystyle\int_{-1/2}^{0} f(t)\,dt + \int_{0}^{x} f(t)\,dt = \frac{\sqrt{3}}{8} + \frac{\pi}{12} + \frac{1}{2}x\sqrt{1-x^2} + \frac{1}{2}\arcsin x$

     (b) Yes.

19. (a) $A_f(x) = x^3/3$

     (b) Yes.

5.3 The fundamental theorem of calculus

1. (a) Graph A is the winner. The point is that $F' = f$. Graph C is wrong because it goes down in the middle. Graph B has the wrong direction of concavity.

   (b) The g-graph is just like the F-graph, but *raised* two units vertically. As always in this section, the point is that F and g have the same derivative, f.

3. (a) $F'(x) = f(x)$. Therefore, $F'(x) > 0$ when $0 < x < 1$, $3 < x < 5$, and $7 < x < 9$; $F'(x) < 0$ when $1 < x < 3$, $5 < x < 7$, and $9 < x \leq 10$. Thus, $F(x)$ has local maxima at $x = 1$, $x = 5$, and $x = 9$.

   (b) $F(x)$ attains its minimum at $x = 3$.

   (c) $f'(x) = F''(x) > 0$ when $2 < x < 4$, and $6 < x < 8$. Therefore, the graph of F is concave up on these intervals.

5. (a) $\displaystyle\int_1^4 \left(x + x^{3/2} \right) dx = 199/10$.

   (b) $\displaystyle\int_0^\pi \cos x \, dx = 0$

   (c) $\displaystyle\int_{-2}^5 \frac{dx}{x+3} = \ln 8$

   (d) $\displaystyle\int_0^b x^2 \, dx = b^3/3$

   (e) $\displaystyle\int_1^b x^n \, dx = \frac{b^{n+1}}{n+1} - \frac{1}{n+1} \quad [n \neq -1]$

   (f) $\displaystyle\int_2^{2.001} \frac{x^5}{1000} \, dx = \frac{2.001^6 - 2^6}{6000} \approx 0.00003204$

   (g) $\displaystyle\int_0^{0.001} \frac{\cos x}{1000} \, dx = \frac{\sin 0.001}{1000} \approx 0.000001$

   (h) $\displaystyle\int_0^{\sqrt{\pi}} x \sin \left(x^2 \right) dx = -\frac{1}{2} \cos(x^2) \Big]_0^{\sqrt{\pi}} = 1$

7. By the fundamental theorem, $F'(x) = \sqrt[3]{x^2 + 7}$, so $F'(1) = \sqrt[3]{8} = 2$. Thus the slope of the tangent line is 2; since the graph of F passes through the point $(1, 0)$, $y = 2x - 2$ is an equation for the tangent line.

9. (a) $\int_{-1}^1 f(x) \, dx = \int_{-1}^0 f(x) \, dx + \int_0^1 f(x) \, dx = -2 + 1 = -1$

   (b) $F(x) = \int_{-2}^x f(t) \, dt$. So when $x < 0$, $F(x) = -(2x + 4)$ and when $x \geq 0$, $F(x) = x - 4$. Here's the picture:

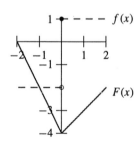

   (c) $F'(1) = f(1) = 1$

   (d) $F'(-1) = f(-1) = -2$

   (e) $F'(0)$ does not exist because $F'(x) = 1$ for all $x > 0$ and $F'(x) = -2$ for all $x < 0$. At $x = 0$, F has a sharp bend.

11. (a) Let $F(x) = \int_1^x \frac{dt}{t}$. Then, using the chain rule, $\dfrac{d}{dx} F(ax) = aF'(ax) = a \cdot \dfrac{1}{ax} = \dfrac{1}{x}$.

    (b) Since $\ln(ax)$ has the same derivative as $\ln x$, it differs from that function only by a constant. Thus, $\ln(ax) = \ln x + C$.

    (c) Let $x = 1$. Then $\ln(ax) = \ln x + C \implies \ln(a) = C$ (since $\ln 1 = 0$).

5.4   Approximating sums: the integral as a limit

1. Using 5 equal subintervals, the left sum approximation to $\int_{-5}^{5} g(x)\,dx$ is

$$2\Big(f(-5) + f(-3) + f(-1) + f(1) + f(3)\Big) = 4;$$

the right sum approximation is $2\Big(f(-3) + f(-1) + f(1) + f(3) + f(5)\Big) = 8$;

and, the midpoint sum approximation is $2\Big(f(-4) + f(-2) + f(0) + f(2) + f(4)\Big) = 7$.

Diagrams illustrating each of these sums appear below:

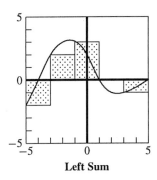   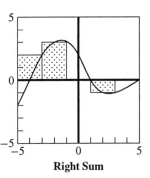

Left Sum   **Right Sum**

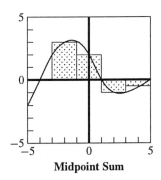

Midpoint Sum

3. left: $\displaystyle\int_{0}^{5} \sqrt[3]{2x}\,dx \approx \frac{5}{10}\sum_{j=0}^{9}\sqrt[3]{2\cdot j\cdot\frac{5}{10}} = \frac{1}{2}\sum_{j=0}^{9}\sqrt[3]{j}$

   right: $\displaystyle\int_{0}^{5} \sqrt[3]{2x}\,dx \approx \frac{5}{10}\sum_{j=1}^{10}\sqrt[3]{2\cdot j\cdot\frac{5}{10}} = \frac{1}{2}\sum_{j=1}^{10}\sqrt[3]{j}$

   midpoint: $\displaystyle\int_{0}^{5} \sqrt[3]{2x}\,dx \approx \frac{5}{10}\sum_{j=0}^{9}\sqrt[3]{2\cdot(j+0.5)\cdot\frac{5}{10}} = \frac{1}{2}\sum_{j=0}^{9}\sqrt[3]{j+0.5}$

5. Let $S = \dfrac{1}{10}\displaystyle\sum_{k=1}^{40}\cos\left(\dfrac{2k-1}{10}\right) = \dfrac{1}{10}\big(\cos(1/10) + \cos(3/10) + \cos(5/10) + \cdots + \cos(79/10)\big)$. Staring reveals that each
   of the subintervals has length $2/10 = 1/5$; their endpoints are $0, 2/10, 4/10, 6/10, \ldots, 78/10, 8$. Writing the sum S in the
   form

$$
\begin{aligned}
S &= \frac{1}{10}\big(\cos(1/10) + \cos(3/10) + \cos(5/10) + \cdots + \cos(79/10)\big) \\
  &= \frac{2}{10}\big(\tfrac{1}{2}\cos(1/10) + \tfrac{1}{2}\cos(3/10) + \tfrac{1}{2}\cos(5/10) + \cdots + \tfrac{1}{2}\cos(79/10)\big)
\end{aligned}
$$

shows that S is the midpoint approximation M_{40} for the integral $\int_0^8 \frac{1}{2}\cos(x)\,dx$. This integral is easy to evaluate:
$\int_0^8 \frac{1}{2}\cos(x)\,dx = \frac{1}{2}\sin(x)\Big]_0^8 = \frac{1}{2}\sin 8 \approx 0.49468$.

7.   (a) $\dfrac{24}{N}\displaystyle\sum_{j=0}^{N-1} c(t_j)\cdot E(t_j)$ where $t_j = 24j/N$.   (b) $\displaystyle\int_0^{24} c(t)E(t)\,dt$

9.   (a) $\displaystyle\int_0^5 r(t)\,dt$

    (b) $\displaystyle\int_0^5 r(t)\,dt \approx 32e^{0.05} + 32e^{0.1} + 32e^{0.15} + 32e^{0.2} + 32e^{0.25}$

    (c) Each term corresponds approximates the amount of oil consumed in one year.

    (d) $\displaystyle\int_0^5 32e^{0.05t}\,dt = \dfrac{32}{0.05}e^{0.05t}\Big]_0^5 = 640\left(e^{0.25} - 1\right) \approx 181.78$

11. $\displaystyle\lim_{n\to\infty} \frac{2}{n}\sum_{j=1}^{n}\left(\frac{2j}{n}\right)^3 = \int_0^2 x^3\,dx = 4$

13. We leave pictures to you; here are the numerical answers.

    (a) $L_4 = 6$, $R_4 = 6$, $M_4 = 5$, $T_4 = 6$; exact answer is $16/3$.

    (b) $L_4 = -8$, $R_4 = 8$, $M_4 = 0$, $T_4 = 0$; exact answer is 0.

    (c) $L_4 = 2$, $R_4 = 6$, $M_4 = 4$, $T_4 = 4$; exact answer is 4.

    (d) $L_4 \approx -0.785$, $R_4 = 0.785$, $M_4 = 0$, $T_4 = 0$; exact answer is 0

5.5   Approximating sums: interpretations and applications

1. (b) The area is approximately $\dfrac{1}{5} \sum_{i=0}^{4} \left(\dfrac{i}{5} - \dfrac{i^2}{25} \right) = \dfrac{5}{32} = 0.15625$.

    (c) Here's the picture:

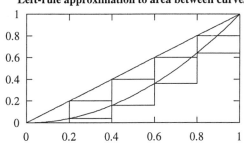

Left-rule approximation to area between curves

    (d) The area is $\displaystyle\int_0^1 (x - x^2)\,dx = \dfrac{1}{2}x^2 - \dfrac{1}{3}x^3 \Big]_0^1 = \dfrac{1}{6}$.

3. Area $= \displaystyle\int_{-1}^{1} (1 - x^4)\,dx = \dfrac{8}{5}$.

4. Area $= \displaystyle\int_{-1}^{0} (x^3 - x)\,dx + \int_0^1 (x - x^3)\,dx = 2\int_0^1 (x - x^3)\,dx = \dfrac{1}{2}$

5. Area $= \displaystyle\int_0^1 (x^2 - x^3)\,dx = \dfrac{1}{12}$

6. Area $= \displaystyle\int_{-1}^{2} [(x + 1) - (x^2 - 1)]\,dx = \dfrac{9}{2}$

7. Area $= \displaystyle\int_{-3}^{2} [(2 - y) - (y^2 - 4)]\,dy = \int_{-4}^{0} 2\sqrt{x+4}\,dx + \int_0^5 \left[(2 - x) + \sqrt{x+4}\right]\,dx = \dfrac{125}{6}$

8. Area $= \displaystyle\int_0^4 \sqrt{x}\,dx = \dfrac{16}{3}$

9. Area $= \displaystyle\int_0^1 (\sqrt{x} - x^2)\,dx = \dfrac{1}{3}$

10. Area $= \displaystyle\int_{-\frac{1}{4}}^{1} \left[(2 - x) - \dfrac{9}{4x+5}\right]\,dx = \dfrac{65}{32} - \dfrac{9}{4}\ln(9) + \dfrac{9}{4}\ln(4) \approx 0.20666$

11. Area $= \displaystyle\int_{-1}^{1} \left[(2 - x^2) - \dfrac{9}{4x^2+5}\right]\,dx = \dfrac{10}{3} - \dfrac{9}{\sqrt{5}}\arctan\left(\dfrac{2}{\sqrt{5}}\right) \approx 0.39624$

12. Area $= \displaystyle\int_1^3 (\sin x - \cos x)\,dx = \cos 1 + \sin 1 - \cos 3 - \sin 3 \approx 2.2306$

13. Area $= \displaystyle\int_{-\frac{\pi}{4}}^{\frac{\pi}{6}} [(2 + \cos x) - \sec^2 x]\,dx \approx 2.2478$

14. Area $= \displaystyle\int_0^1 e^x\,dx = e - 1$

15. Area $= \displaystyle\int_{-1}^{0} (2^x - 5^x)\, dx + \int_{0}^{1} (5^x - 2^x)\, dx = \frac{16}{5\ln 5} - \frac{1}{2\ln 2} \approx 1.2669$

6.1   Introduction

1. $\int \left(3x^2 + 4x^{-2}\right) dx = x^3 - 4/x + C$

3. $\int \left(2\sin(3x) - 4\cos(5x)\right) dx = -\frac{2}{3}\cos(3x) - \frac{4}{5}\sin(5x)$

5. $\int \dfrac{dx}{1 + 4x^2} = \frac{1}{2}\arctan(2x) + C$

7. $\int \dfrac{6}{9 + x^2}\, dx = 2\arctan(x/3) + C$

9. (a) $\displaystyle\int \sin^2 x\, dx \;=\; \int \frac{1}{2}(1 - \cos(2x))\, dx$

$\qquad\qquad\qquad =\; \frac{1}{2}\left(\int 1\, dx - \int \cos(2x)\, dx\right)$

$\qquad\qquad\qquad =\; \frac{1}{2}\left(x - \frac{1}{2}\sin(2x)\right) + C$

$\qquad\qquad\qquad =\; \frac{x}{2} - \frac{1}{4}\sin(2x) + C$

   (b) $\displaystyle\int \cos^2 x\, dx \;=\; \int (1 - \sin^2 x)\, dx$

$\qquad\qquad\qquad =\; \int 1\, dx - \int \sin^2 x\, dx$

$\qquad\qquad\qquad =\; x - \left(\frac{x}{2} - \frac{1}{4}\sin(2x)\right) + C$

$\qquad\qquad\qquad =\; \frac{x}{2} + \frac{1}{4}\sin(2x) + C$

   (c) $\sqrt{1 - \cos(2x)} \neq \sqrt{2}\sin x$ when $\pi < x < 2\pi$.

11. $\left(\ln|\sec x| + C\right)' = \dfrac{\sec x \tan x}{\sec x} = \tan x$

13. $\left(\dfrac{1}{2}\ln\left|\dfrac{1 + x}{1 - x}\right| + C\right)' = \dfrac{1}{2}\left(\dfrac{1}{1 + x} + \dfrac{1}{1 - x}\right) = \dfrac{1}{1 - x^2}$

   Alternatively, $\left(\dfrac{1}{2}\ln\left|\dfrac{1 + x}{1 - x}\right| + C\right)' = \dfrac{1}{2} \cdot \dfrac{1 - x}{1 + x} \cdot \dfrac{2}{(1 - x)^2} = \dfrac{1}{1 - x^2}$

15. $\left(x\ln x - x + C\right)' = \ln x + \dfrac{x}{x} - 1 = \ln x$

17. $\left(-\dfrac{1}{6}(2x + 3)^{-3} + C\right)' = -\dfrac{1}{6} \cdot (-3) \cdot (2x + 3)^{-4} \cdot 2 = (2x + 3)^{-4}$

19. $\dfrac{1}{2}\cos x \sin x = \dfrac{2\cos x \sin x}{4} = \dfrac{1}{4}\sin(2x)$

21. **True.** $\left(\arctan(2x) + C\right)' = \dfrac{2}{1 + 4x^2}$

6.2 Antidifferentiation by substitution

1. $\displaystyle\int (4x+3)^{-3}\,dx = -\frac{1}{8\,(4x+3)^2} + C$

3. $\displaystyle\int e^{\sin x}\cos x\,dx = e^{\sin x} + C$

5. $\displaystyle\int \frac{\arctan x}{1+x^2}\,dx = \frac{1}{2}(\arctan x)^2$

7. $\displaystyle\int \frac{e^{1/x}}{x^2}\,dx = \int \frac{e^{x^{-1}}}{x^2}\,dx = -e^{x^{-1}} + C = -e^{1/x} + C$

9. $\displaystyle\int \frac{e^x}{1+e^{2x}}\,dx = \arctan\left(e^x\right) + C$

11. $\displaystyle\left(\frac{(2x+3)^{3/2}}{6} - \frac{3\sqrt{2x+3}}{2} + C\right)' = \frac{\sqrt{2x+3}}{2} - \frac{3}{2\sqrt{2x+3}} = \frac{x}{\sqrt{2x+3}}.$

13. $u = 3x^2,\ a = 3\pi/2,\ b = 3\pi;\quad \displaystyle\int_{-\sqrt{\pi/2}}^{\sqrt{\pi}} x\cos\left(3x^2\right)dx = \frac{1}{6}\int_{3\pi/2}^{3\pi}\cos u\,du = \frac{1}{6}$

15. $u = x^2,\ a = 4,\ b = 1;\quad \displaystyle\int_{-2}^{1}\frac{x}{1+x^4}\,dx = \frac{1}{2}\int_{4}^{1}\frac{du}{1+u^2} = \tfrac{1}{2}(\pi/4 - \arctan 4) \approx -0.27021$

17. Let $u = 2x+3$. Then, $\displaystyle\int \cos(2x+3)\,dx = \frac{1}{2}\int \cos u\,du = \frac{1}{2}\sin u + C = \frac{1}{2}\sin(2x+3) + C.$

19. Let $u = 3x+2$. Then, $\displaystyle\int x\,(3x+2)^4\,dx \to \frac{1}{9}(u-2)u^4\,du = \frac{1}{54}u^6 - \frac{2}{45}u^5 + C \to \frac{1}{54}(3x+2)^6 - \frac{2}{45}(3x+2)^5 + C.$

21. Let $u = \cos x$. Then, $\displaystyle\int \tan x\,dx = -\ln|\cos x| + C.$

23. Let $u = x^4 - 1$. Then, $\displaystyle\int x^3\left(x^4-1\right)^2 dx = \frac{1}{12}\left(x^4-1\right)^3 + C.$

25. Let $u = x^2$. Then, $\displaystyle\int \frac{x}{\sqrt{1-x^4}}\,dx = \frac{1}{2}\arcsin(x^2) + C.$

27. Let $u = 1+x^2$. Then, $\displaystyle\int \frac{x}{\sqrt{1+x^2}}\,dx = \sqrt{1+x^2} + C.$

29. Let $u = 1-x^2$. Then, $\displaystyle\int x\left(1-x^2\right)^{15} dx = -\frac{1}{32}\left(1-x^2\right)^{16} + C.$

31. Let $u = x^2 + 3x + 5$. Then, $\displaystyle\int \frac{2x+3}{\left(x^2+3x+5\right)^4}\,dx = -\frac{1}{3}\left(x^2+3x+5\right)^{-3} + C.$

33. Let $u = 1-x^2$. Then, $\displaystyle\int x\cos\left(1-x^2\right) dx = -\frac{1}{2}\sin\left(1-x^2\right) + C.$

35. Let $u = 2e^x + 3$. Then, $\displaystyle\int \frac{e^x}{(2e^x+3)^2}\,dx = -\frac{1}{2}\left(2e^x+3\right)^{-1} + C.$

37. Let $u = x+1$. Then, $\displaystyle\int \frac{2x+3}{(x+1)^2}\,dx \to \int \frac{2(u-1)+3}{u^2}\,du \to 2\ln(x+1) - (x+1)^{-1} + C.$

39. Let $u = \sec x$. Then, $\displaystyle\int \sec x \tan x \, dx \to \int du = u \to \sec x + C.$

41. Let $u = \ln x$. Then, $\displaystyle\int \frac{\ln x}{x} \, dx = \frac{1}{2}(\ln x)^2 + C.$

43. Let $u = 1 + \sec x$. Then, $\displaystyle\int \sec x \tan x \sqrt{1 + \sec x} \, dx = \frac{2}{3}(1 + \sec x)^{3/2} + C.$

45. Let $u = \sin x$. Then, $\displaystyle\int \frac{\cos x}{1 + \sin^2 x} \, dx = \arctan(\sin x) + C.$

47. Let $u = \tan x$. Then, $\displaystyle\int \sec^2 \tan x \, dx = \frac{1}{2}\tan^2 x + C.$

49. Let $u = x^3$. Then, $\displaystyle\int x^2 \sec^2\left(x^3\right) dx = \frac{1}{3}\tan\left(x^3\right) + C.$

51. Let $u = x^5 + 6$. Then, $\displaystyle\int x^4 \sqrt[3]{x^5 + 6} \, dx = \frac{3}{20}\left(x^5 + 6\right)^{4/3} + C.$

53. Let $u = 1 + x^2$. Then, $\displaystyle\int_0^2 \frac{x}{\left(1 + x^2\right)^3} \, dx = -\frac{1}{4}\left(1 + x^2\right)^{-2}\Bigg]_0^2 = \frac{6}{25}$

55. Let $u = \ln x$. Then, $\displaystyle\int_1^e \frac{\sin(\ln x)}{x} \, dx = -\cos(\ln x)\Bigg]_1^e = 1 - \cos 1$

57. Let $u = \sin x$. Then, $\displaystyle\int_0^\pi \sin^3 x \cos x \, dx = \frac{1}{4}\sin^4 x\Bigg]_0^\pi = 0$

59. (a) When $u = \sec x$, $du = \sec x \tan x \, dx$ so

$$I = \int \sec^2 x \tan x \, dx = \int \sec x \cdot \sec x \tan x \, dx \to \int u \, du = \frac{1}{2}u^2 + C \to \frac{1}{2}\sec^2 x + C.$$

(b) When $u = \tan x$, $du = \sec^2 x \, dx$ so

$$I = \int \sec^2 x \tan x \, dx = \int \tan x \cdot \sec^2 x \, dx \to \int u \, du = \frac{1}{2}u^2 + C \to \frac{1}{2}\tan^2 x + C.$$

(c) Since $\sec^2 x = 1 + \tan^2 x$, the expressions in parts (a) and (b) differ from each other by a constant (1/2). Because an antiderivative is unique only up to an additive constant (i.e., two antiderivatives of the same function may differ by a constant), there is no paradox.

61. Let $u^2 = 2x + 1$. Then, $dx = u \, du$ and

$$\int x\sqrt{2x + 1} \, dx = \frac{1}{2}\int (u^2 - 1)u^2 \, du = \frac{1}{2}\left(\frac{u^5}{5} - \frac{u^3}{3}\right) + C = \frac{1}{10}(2x + 1)^{5/2} - \frac{1}{6}(2x + 1)^{3/2} + C.$$

6.3 Integral aids: tables and computers

1. $\int \dfrac{dx}{3 + 2e^{5x}} = \dfrac{1}{3}x - \dfrac{1}{15}\ln\left(3 + 2e^{5x}\right) + C.$ [Use formula #60.]

3. $\int \dfrac{dx}{x^2(3 - x)} = -\dfrac{1}{3x} - \dfrac{1}{9}\ln\left|\dfrac{x - 3}{x}\right| + C = \dfrac{1}{9}\ln\left|\dfrac{x}{x - 3}\right| - \dfrac{1}{3x} + C.$ [Use formula #24.]

5. $\int \tan^3(5x)\,dx = \dfrac{1}{10}\tan^2(5x) + \dfrac{1}{5}\ln|\cos(5x)| + C.$ [Use formulas #52 and #7.]

7. $\int x\sin(2x)\,dx = \dfrac{1}{4}\sin(2x) - \dfrac{x}{2}\cos(2x) + C.$ [Use formula #48.]

9. $\int e^{2x}\cos(3x)\,dx = \dfrac{e^{2x}}{13}\left(2\cos(3x) + 3\sin(3x)\right) + C.$ [Use formula #57.]

11. $\int \dfrac{dx}{4 - x^2}\,dx = \dfrac{1}{4}\ln\left|\dfrac{2 + x}{2 - x}\right| + C.$ [Use formula #32.]

13. $\int \dfrac{4x + 5}{(2x + 3)^2}\,dx = 4\int \dfrac{x}{(2x + 3)^2}\,dx + 5\int \dfrac{dx}{(2x + 3)^2} = \dfrac{1}{2(2x + 3)} + \ln|2x + 3| + C.$
[Use formulas #22 and #19.]

15. $\int \dfrac{dx}{4x^2 - 1} = \dfrac{1}{4}\ln\left|\dfrac{2x - 1}{2x + 1}\right| + C.$ [Use formula #34.]

17. $\int \dfrac{x + 2}{2 + x^2}\,dx = \int \dfrac{x}{2 + x^2}\,dx + 2\int \dfrac{dx}{2 + x^2} = \dfrac{1}{2}\ln\left(x^2 + 2\right) + \sqrt{2}\arctan(x\sqrt{2}/2) + C.$
[Use the substitution $u = 2 + x^2$ to evaluate one antiderivative and formula #33 to evaluate the other.]

19. $\int \dfrac{5}{4x^2 + 20x + 16}\,dx = \dfrac{5}{4}\int \dfrac{dx}{x^2 + 5x + 4}\,dx = \dfrac{5}{4}\int \dfrac{dx}{(x + 5/2)^2 - 9/4}\,dx = \dfrac{5}{12}\ln\left|\dfrac{x + 1}{x + 4}\right| + C.$
[Use formula #34.]

21. $\int x^3\cos(x^2)\,dx \to \dfrac{1}{2}\int u\cos u\,du = \dfrac{1}{2}(u\sin u + \cos u) + C \to \dfrac{1}{2}\left(x^2\sin\left(x^2\right) + \cos\left(x^2\right)\right) + C.$
[Make the substitution $u = x^2$ in formula #49.]

23. $\int \dfrac{dx}{(x^2 + 3x + 2)^2} = \int \dfrac{dx}{\left((x + 3/2)^2 - 1/4\right)^2} = -\dfrac{2x + 3}{x^2 + 3x + 2} - 2\ln\left|\dfrac{x + 1}{x + 2}\right| + C.$

25. $\int \dfrac{e^x}{e^{2x} - 2e^x + 5}\,dx \to \int \dfrac{du}{u^2 - 2u + 5} = \int \dfrac{du}{(u - 1)^2 + 4} \to \int \dfrac{dw}{w^2 + 4} =$
$\dfrac{1}{2}\arctan(w/2) + C \to \dfrac{1}{2}\arctan\left(\dfrac{e^x - 1}{2}\right) + C.$

27. $\int \sqrt{x^2 + 4x + 1}\,dx = \int \sqrt{(x + 2)^2 - 3}\,dx = \dfrac{1}{2}\left((x + 2)\sqrt{x^2 + 4x + 1} - 3\ln\left|x + 2 + \sqrt{x^2 + 4x + 1}\right|\right) + C.$
[Complete the square, then use formula #38.]

29. $\displaystyle \int \frac{\cos x \sin x}{(\cos x - 4)(3\cos x + 1)}\, dx \;=\; \int \frac{\cos x \sin x}{3\cos^2 x - 11\cos x - 4}\, dx$

$\displaystyle \rightarrow \; -\int \frac{u}{3u^2 - 11u - 4}\, du$

$\displaystyle = \; -\frac{1}{3}\int \frac{u}{u^2 - 11u/3 - 4/3}\, du = -\frac{1}{3}\int \frac{u}{(u - 11/6)^2 - 169/36}\, du$

$\displaystyle \rightarrow \; -\frac{1}{3}\int \frac{w + 11/6}{w^2 - (13/6)^2}\, dw$

$\displaystyle = \; -\frac{4}{13}\ln|w - 13/6| - \frac{1}{39}\ln|w + 13/6| + C$

$\displaystyle \rightarrow \; -\frac{4}{13}\ln|u - 4| - \frac{1}{39}\ln|u + 1/3| + C$

$\displaystyle \rightarrow \; -\frac{4}{13}\ln|\cos x - 4| - \frac{1}{39}\ln|\cos x + 1/3| + C$

31. $\displaystyle \int x \sin(3x + 4)\, dx \;\rightarrow\; \frac{1}{9}\int (u - 4)\sin u\, du$

$\displaystyle = \; \tfrac{1}{9}\left(\sin u - u\cos u + 4\cos u\right) + C$

$\displaystyle \rightarrow \; \tfrac{1}{9}\left(\sin(3x + 4) - (3x + 4)\cos(3x + 4) + 4\cos(3x + 4)\right) + C$

7.1   The idea of approximation

1.   (a) $I = \dfrac{x^4}{4}\bigg]_0^1 = \dfrac{1}{4} = \dfrac{16}{64}$.

   (b) $L_4 = (f(0) + f(1/4) + f(1/2) + f(3/4)) \cdot \dfrac{1}{4} = \left(0^3 + (1/4)^3 + (1/2)^3 + (3/4)^3\right) \cdot \dfrac{1}{4} = \dfrac{9}{64}$.

   Similarly, $R_4 = (f(1/4) + f(1/2) + f(3/4) + f(1)) \cdot \dfrac{1}{4} = \left((1/4)^3 + (1/2)^3 + (3/4)^3 + 1^3\right) \cdot \dfrac{1}{4} = \dfrac{25}{64}$.

   (c) The error bounds are the same for both L_4 and R_4:

$$|I - L_4| \le \frac{|f(1) - f(0)| \cdot (1 - 0)}{4} = \frac{1}{4}.$$

   The actual errors are less: 7/64 and 9/64, respectively.

   (d) $T_4 = \dfrac{L_4 + R_4}{2} = \dfrac{17}{64}$; the actual error is only 1/64—much less than what the theorem predicts.

   (e) For L_n and R_n, the theorem says we're OK if $\dfrac{|f(1) - f(0)| \cdot (1 - 0)}{n} = \dfrac{1}{n} < 0.005$, i.e., if $n > 200$.

3. For $I = \int_1^3 x\,dx$, $f(x) = x$, $a = 1$, $b = 3$, so

$$|f(b) - f(a)|\frac{(b-a)}{n} = 2 \cdot \frac{2}{n} = \frac{4}{n} \le 0.000005 \iff n \ge 800{,}000.$$

5. For $\int_1^4 \sqrt{x}\,dx$ we need $n \ge 600{,}000$.

7. For $\int_2^3 \sin x\,dx$ we need $n \ge 153{,}640$.

13–17. The answers for T_n are, in each case, *half* of those for L_n and R_n: (a) $n \ge 400{,}000$; (b) $n \ge 300{,}000$; (c) $n \ge 300{,}000$; (d) $n \ge 50{,}000$; (e) $n \ge 76{,}820$.

15.   (a) $I = \displaystyle\int_0^\pi \sin x\,dx = -\cos x\bigg]_0^1 = 2$.

   (b) $|I - L_4| = |2 - \pi(\sqrt{2}+1)| \approx 1.8961$.

   (c) Theorem 1 doesn't apply here, because the integrand is *not* monotone.

17. For a *decreasing* function: $R_n \le M_n \le L_n$. (A picture should make this clear.)

19. The key idea is that for equally-spaced points $a = x_0 < x_1 < x_2 < \cdots < x_n = b$,

$$L_n = f(x_0)\Delta x + f(x_1)\Delta x + f(x_2)\Delta x + \cdots + f(x_{n-1})\Delta x,$$

while

$$R_n = f(x_1)\Delta x + f(x_1)\Delta x + f(x_2)\Delta x + \cdots + f(x_n)\Delta x,$$

where $\Delta x = (b - a)/n$. Thus, by algebra,

$$R_n - L_n = f(x_n)\Delta x - f(x_0)\Delta x = \big(f(b) - f(a)\big)\Delta x = \big(f(b) - f(a)\big)\frac{b-a}{n}$$

or, equivalently, $R_n = L_n + \big[f(b) - f(a)\big] \cdot \dfrac{(b-a)}{n}$.

21. If $f(a) = f(b)$, then $f(b) - f(a) = 0$. In this case, $R_n = L_n$ (see problem 15), so $T_n = L_n$ (see problem 16).

23. As the picture in the proof of Theorem 1 shows, the "exact" integral I must lie somewhere *between* L_n and R_n, i.e., in an interval of length

$$|R_n - L_n| = |f(b) - f(a)|\Delta x = |f(b) - f(a)|\frac{(b-a)}{n}.$$

Since T_n is the *midpoint* of this same interval, I must lie within a distance of *half* the interval's width from T_n. (Draw L_n, R_n, and T_n on a number line to understand all this.)

25. Notice that the integrand is *not* monotone over the interval $[0, 2]$. Thus before using Theorem 1 we should break the interval of integration two pieces, on each of which the integrand *is* monotone. A look at the graph shows that the function is *increasing* on $[0, \sqrt{\pi/2}]$ and *decreasing* on $[\sqrt{\pi/2}, 2]$. Thus we can approximate answers on each of these intervals, say with error less than 0.005, and add the results. The respective answers are about

$$\int_0^{\sqrt{\pi/2}} \sin\left(x^2\right) dx \approx 0.549276; \qquad \int_{\sqrt{\pi/2}}^2 \sin\left(x^2\right) dx \approx 0.255652;$$

a good estimate to I, therefore, is around 0.80 or 0.81. (More advanced methods can be used to show that the "true" answer is about 0.80485.)

27. The point of (iv), in each case, is to see that the ACTUAL errors committed are no more than the theoretical bounds guaranteed by Theorem 1. Each problem has many parts; here they are, tabulated.

(a) $I = \int_1^2 x^2 \, dx = 7/3$.

n	4	8	16	32	64	128
L_n	1.9688	2.1485	2.2403	2.2867	2.3103	2.3220
$I - L_n$	0.3645	0.1848	0.0930	0.0466	0.0230	0.0113
Thm 1 bound	0.7500	0.3750	0.1875	0.0937	0.0468	0.0234

(b) $I = \int_1^4 \sqrt{x} \, dx = 14/3$.

n	4	8	16	32	64	128
L_n	4.2802	4.4764	4.5724	4.6199	4.6435	4.6560
$I - L_n$	0.3865	0.1903	0.0943	0.0468	0.0232	0.0107
Thm 1 bound	0.75000	0.37500	0.18750	0.0937	0.0468	0.0234

(c) $I = \int_1^2 x^{-1} \, dx = \ln 2 \approx 0.6932$.

n	4	8	16	32	64	128
L_n	0.75953	0.72538	0.70900	0.70103	0.69709	0.69514
$I - L_n$	−0.0664	−0.0322	−0.01585	−0.0079	−0.0039	−0.0020
Thm 1 bound	−0.1250	−0.0625	−0.0312	−0.0156	−0.0078	−0.0039

(d) $I = \int_2^3 \sin x \, dx = -\cos 3 + \cos 2 \approx 0.57384$.

n	4	8	16	32	64	128
L_n	0.6669	0.6211	0.5976	0.5858	0.5798	0.5768
$I - L_n$	−0.0930	−0.0472	−0.0238	−0.0120	−0.0600	−0.0301
Thm 1 bound	−0.19205	−0.09602	−0.04801	−0.02400	−0.01200	−0.00600

29. The answers can be read from the tables found in Problem 31. The point is to see that for the Left rule, *doubling n* roughly *halves* the error committed. Thus the error committed with $n = 512$ should be roughly 1/4 that committed with $n = 128$.

7.2   More on error: left and right sums and the first derivative

The following table contains all the numerical information. Numbers are rounded to three or four decimal places. Notice that in each case the value of K_1 is somewhat arbitrary—larger values of K_1 would also work.

Problem	I	L_{10}	R_{10}	$|I - L_{10}|$	$|I - R_{10}|$	K_1	$K_1(b-a)^2/2n$
1. $\int_2^3 1\,dx$	1.	1.000	1.000	0	0	0	0
2. $\int_1^3 x\,dx$	4.	3.800	4.200	0.200	0.200	1	0.2000
3. $\int_1^2 x^2\,dx$	2.333	2.185	2.485	0.148	0.152	4	0.2000
4. $\int_1^4 \sqrt{x}\,dx$	4.667	4.512	4.812	0.155	0.145	0.5	0.2250
5. $\int_1^2 x^{-1}\,dx$	0.6931	0.7187	0.6687	0.0256	0.0244	1	0.0500
6. $\int_2^3 \sin x\,dx$	0.5739	0.6120	0.5351	0.0381	0.0388	1	0.0500

7. For $\int_0^3 e^{-x^2}\,dx$ we consider $f'(x) = -\exp(-x^2)2x$. On $[0, 3]$, $|f'(x)| \le 0.86$, so $K_1 = 0.86$ works. Hence we need $n \ge 0.86 \cdot 3^2 \cdot 100 = 774$.

9. For $\int_0^1 \left(1 + x^2\right)^{-1}\,dx$, $K_1 = 0.7$ works, so $n \ge 0.7 \cdot 1^2 \cdot 100 = 70$.

11. (a) $I = \pi$ because the integral gives the area of the northeast quadrant of a circle of radius 2.

    (b) $L_{10} \approx 3.3045;\ \pi - L_{10} \approx 0.1629$.

    (c) Theorem 2 doesn't give a good bound here because we can't compute K_1—$f'(x)$ is unbounded on the interval $(0, 2)$.

    (d) Theorem 1 says that

$$|L_{10} - I| \le \frac{|f(2) - f(0)| \cdot 2}{10} = 0.4.$$

    This number is larger, as it should be, than the actual error committed.

13. The information given implies that f is increasing and concave down on the interval $[a, b]$. A sketch shows that the approximation error made by L_n includes all of the area corresponding to the approximation error made by T_n and more.

    An algebraic proof of this result is also possible. Since f is (strictly) increasing and (strictly) concave down on the interval of integration, $L_n < I < R_n$ and $I - T_n < 0$. Thus,

$$(I - L_n) - (I - T_n) = T_n - L_n = \tfrac{1}{2}\left(R_n - L_n\right) > 0$$

which implies that $|I - T_n| < |I - L_n|$.

15. Any linear function (i.e., of the form $f(x) = ax + b$) will do.

17. We'll tabulate a lot of information first:

Problem	I	T_{10}	M_{10}	$I - T_{10}$	$I - M_{10}$
1. $\int_2^3 1\,dx$	1.	1.0000	1.0000	0	0
2. $\int_1^3 x\,dx$	4.	4.0000	4.0000	0	0
3. $\int_1^2 x^2\,dx$	2.3333	2.3350	2.3327	−0.0017	0.0006
4. $\int_1^4 \sqrt{x}\,dx$	4.6667	4.6650	4.6677	0.0017	−0.0010
5. $\int_1^2 x^{-1}\,dx$	0.69315	0.69378	0.69284	−0.00063	0.00031
6. $\int_2^3 \sin x\,dx$	0.57384	0.57339	0.57410	0.00045	−0.00026

    (a) The approximation errors made by T_{10} and M_{10} are generally considerably less than those made by L_{10} and R_{10}.

(b) Approximation errors made by T_{10} are generally about twice the size of, and opposite in sign to, those made by M_{10}.

(c) T_{10} makes no approximation error on the first two—in both cases, the first derivative is constant.

(d) T_{10} underestimates when the integrand is concave down, i.e., for integrands with *negative* second derivative.

(e) T_{10} overestimates when the integrand is concave up, i.e., for integrands with *positive* second derivative.

(f) M_{10} makes no approximation error on the first two—in both cases, the first derivative is constant.

(g) M_{10} underestimates when the integrand is concave up, i.e., for integrands with *positive* second derivative.

(h) M_{10} overestimates when the integrand is concave down, i.e., for integrands with *negative* second derivative.

19. A picture shows that for any linear function f, T_n commits zero error in approximating $\int_a^b f(x)\,dx$.

21. The statement is true—the larger the derivative is on an interval, the worse L_n behaves there.

7.3   Trapezoid sums, midpoint sums, and the second derivative

1. (a) $M_2 = \frac{1}{2}(f(\frac{1}{4}) + f(\frac{3}{4})) = \frac{e^{1/16} + e^{9/16}}{2} \approx 1.409$.

   $T_2 = \frac{1}{4}(f(0) + 2f(\frac{1}{2}) + f(1)) = \frac{1 + 2e^{1/4} + e^1}{4} \approx 1.571$.

   (b) In all parts below, we'll use $K_1 = 6$, $K_2 = 16.31$. Then

   $L_{10} \approx 1.381$. Error bound: $|L_{10} - I| \leq \dfrac{K_1 \cdot 1}{20} = \dfrac{6}{20} = 0.3$.

   $R_{10} \approx 1.553$. Error bound: same as L_{10}.

   $M_{10} \approx 1.460$. Error bound: $|M_{10} - I| \leq \dfrac{K_2 \cdot 1^3}{24 \cdot 100} = \dfrac{16.31}{2400} \approx 0.007$.

   $T_{10} \approx 1.467$. Error bound: $|T_{10} - I| \leq \dfrac{K_2 \cdot 1^3}{12 \cdot 100} = \dfrac{16.31}{1200} \approx 0.0136$.

   Note that T_{10} and R_{10} *overestimate* the error; the others underestimate.

   (c) We want $|M_n - I| \leq \dfrac{K_2(b-a)^3}{24n^2} = \dfrac{16(1)^3}{24n^2} < 0.0005$. The last inequality holds if $n^2 > \dfrac{16}{24(0.0005)} \approx 1533.33$, i.e.,

   if $n > \sqrt{\dfrac{4000}{3}} \approx 36$. In particular, $n = 37$ is OK, so $M_{37} \approx 1.46248\ldots$; the answer is good to 3 decimal places.

Problem	K_2	I	T_{10}	$\lvert I - T_{10} \rvert$	$\dfrac{K_2(b-a)^3}{12n^2}$	M_{10}	$\lvert I - M_{10} \rvert$	$\dfrac{K_2(b-a)^3}{24n^2}$
3. $\int_2^3 1\,dx$	0	1.	1.0000	0	0	1.0000	0	0
4. $\int_1^3 x\,dx$	0	4.	4.0000	0	0	4.0000	0	0
5. $\int_1^2 x^2\,dx$	2	2.3333	2.3350	0.0017	0.0017	2.3327	0.0006	0.0008
6. $\int_1^4 \sqrt{x}\,dx$	0.25	4.6667	4.6650	0.0017	0.0056	4.6677	0.0001	0.0028
7. $\int_1^2 x^{-1}\,dx$	2	0.6932	0.6938	0.0006	0.0017	0.6928	0.00031	0.0008
8. $\int_2^3 \sin x\,dx$	1	0.57384	0.57339	0.00045	0.00083	0.57410	0.00026	0.00042

9. Let $I = \int_a^b f(x)\,dx$. Since f is increasing on $[a, b]$, the left rule *underestimates* I, the right rule *overestimates*, and the other two lie in between. Since f is concave up on the interval $[a, b]$, $M_n \leq T_n$ must be true. Thus, $L_n \leq M_n \leq T_n \leq R_n$ so $L_n = 8.52974$, $M_n = 9.71090$, $T_n = 9.74890$, and $R_n = 11.04407$.

11. Since $F''(x) = f'(x) \geq 0$ on $[a, b]$, T_{100} overestimates the value of I.

13. (a) **No.** Since f' is an increasing function on the interval of integration, f is concave upwards on this interval. It follows that M_n *under*estimates I (i.e., $M_n < I$).

    (b) Since $F'' = f'$, the graph shows that we can take $K_2 = 5.5$. Thus, any $n \geq 5$ will work.

15. (a) $R_4 = 0.75 \cdot (26.522 + 48.755 + 68.328 + 86.790) = 172.79625$.

    $|I - R_4| \leq 0.75 \cdot |86.790 - 2.0000| = 63.5925$.

    (b) $T_4 = \frac{1}{2}(L_4 + R_4) = \frac{1}{2} \cdot 0.75 \cdot (2 - 86.790) + R_4 = 141.00$.

    From the graph, it is apparent that $\left|f''(x)\right| < 6$ when $-1 \leq x \leq 2$ so, taking $K_2 = 6$, we have

    $$|I - T_4| \leq \frac{6 \cdot 3^3}{12 \cdot 4^2} = 0.84375.$$

The key point in exercises 17–20 is the fact that *quadratic* functions—the ones for which f'' is *constant*—are the "worst offenders" for the trapezoid and midpoint rules. The other ingredient is concavity: the trapezoid rule underestimates if f is concave down and overestimates if f is concave up; the midpoint rule does just the opposite. Hence all parts below amount to choosing a quadratic function with the appropriate concavity.

17. If $f(x) = x^2$, then f is concave up, so M_n underestimates I and the approximation error is as bad as Theorem 3 allows.

19. If $f(x) = -x^2$, then f is concave down, so T_n underestimates I and the approximation error is as bad as Theorem 3 allows.

21. (a) The error bounds for M_n and M_{10n} are

$$|I - M_n| \leq \frac{K_2(b-a)^3}{24n^2}; \qquad |I - M_{10n}| \leq \frac{K_2(b-a)^3}{24 \cdot (10n)^2} = \frac{K_2(b-a)^3}{24 \cdot 100n^2}.$$

The second bound has an extra factor of 100 in the denominator. This means that using ten times as many subintervals in M_n gives about *two* extra decimal places of accuracy.

(b) The error bounds for L_n and L_{10n} are

$$|I - L_n| \leq \frac{K_1(b-a)^2}{2n}; \qquad |I - L_{10n}| \leq \frac{K_1(b-a)^2}{2 \cdot 10n} = \frac{K_1(b-a)^2}{20n}.$$

Here the second bound has an extra factor of 10 in the denominator, using ten times as many subintervals in L_n gives only *one* extra decimal place of accuracy.

7.4 Simpson's rule

1. For $I = \int_a^b dx = x\Big]_a^b = b - a$. We need to show that S_2 has the *same* value as I. Here goes:

$$
\begin{aligned}
S_2 &= \frac{b-a}{6}\big(f(a) + 4f((a+b)/2) + f(b)\big) \\
&= \frac{b-a}{6}(1 + 4 \cdot 1 + 1) \\
&= b - a.
\end{aligned}
$$

3. For $I = \int_a^b x^2\,dx = \frac{x^3}{3}\Big]_a^b = \frac{b^3 - a^3}{3}$. We need to show that S_2 has the *same* value as I. Here goes:

$$
\begin{aligned}
S_2 &= \frac{b-a}{6}\big(f(a) + 4f((a+b)/2) + f(b)\big) \\
&= \frac{b-a}{6}\left(a^2 + 4\frac{(a+b)^2}{4} + b^2\right) \\
&= \frac{b-a}{6}\left(2a^2 + 2ab + 2b^2\right) \\
&= \frac{b-a}{3}\left(a^2 + ab + b^2\right) \\
&= \frac{b^3 - a^3}{3} \quad \text{(by a little algebra).}
\end{aligned}
$$

5. Let $I = \int_a^b x^4\,dx = \frac{b^5}{5} - \frac{a^5}{5}$. By comparison, $S_2 = \frac{b-a}{6}\left(a^4 + 4((b+a)/2)^4 + b^4\right)$.

   Some careful algebra now shows: $|I - S_2| = \left|\frac{b^5 - a^5}{5} - \frac{(b-a)}{6}\left(a^4 + \frac{(b+a)^4}{4} + b^4\right)\right| = \frac{(b-a)^5}{120}$.

   The error bound is: $|S_2 - I| \le \frac{K_4(b-a)^5}{180n^4} = \frac{24(b-a)^5}{180 \cdot 2^4} = \frac{(b-a)^5}{120}$. Thus S_2 does commit the maximum possible error.

7. (a) $I = \int_0^1 \cos(100x)\,dx = \frac{\sin(100x)}{100}\Big]_0^1 = \frac{\sin 100}{100} \approx -0.0050637.$

   (b) $S_{10} \approx 0.036019$; thus the actual approximation error is $|I - S_{10}| \approx 0.041083 < 0.05$.

   (c) From the error bound formula, with $K_4 = 100^4 = 10^8$ and $n = 10$, we get

$$
|error| \le \frac{K_4 \cdot 1^5}{180 \cdot 10^8} = \frac{10^8}{180 \cdot 10^4} \approx 55.556.
$$

   This bound is large because K_4 is so large.

9. $S_4 = \frac{1}{3}(2M_2 + T_2) = 141.425$.

   When $-1 \le x \le 2$, $\left|f^{(4)}(x)\right| \le 8 = K_4$, so $|I - S_4| \le \frac{8 \cdot 2^5}{180 \cdot 4^4} = 0.0421875$.

11. By examining graphs of the first, second, and fourth derivatives of the integrand, we determine that $K_1 \ge 1$, $K_2 \ge 3.25$, and $K_4 \ge 47.5$. Thus, to achieve the desired accuracy, L_n or R_n requires $n \ge 2000$; T_n requires $n \ge 45$; M_n requires $n \ge 33$; and, S_n requires $n \ge 10$. [NOTES: The integrand is not monotonic over the interval of integration, so Theorem 1 cannot be used to produce an error bound. Also, this integral cannot be evaluated using the Fundamental Theorem of Calculus because no elementary antiderivative exists.]

    Sample answers: $M_{33} = 0.45148$, $S_{10} = 0.45131$.

13.   (a) We'll do L_4 explicitly; the others are similar.

$$L_4 = \frac{1}{4}\big(f(0) + f(0.25) + f(0.5) + f(0.75)\big) = 1.1485.$$

Similar calculations show:

$$R_4 = 1.1805; \quad M_2 = 1.1345; \quad T_4 = 1.1645; \quad S_4 = 1.1545.$$

(b) Upward concavity means that T_4 must *overestimate* and M_2 must *underestimate*. Therefore I has to lie in the interval [1.1345, 1.1645]. (The S_4 estimate is consistent with this.)

8.1 Introduction

1. We know that $r_B(t) = A + B(t - 35) + C(t - 35)^2$ for some constants A, B, and C. We'll show that the only possible values for A, B, and C are $A = 13$, $B = 0$, and $C = -1/100$.

Note first that
$$r_B(t) = A + B(t - 35) + C(t - 35)^2 \implies r_B'(t) = B + 2C(t - 35).$$

From the problem we know that $r_B(35) = 13$, $r_B(60) = 27/4$, and $r_B'(35) = 0$. This leads to three simple conditions on A, B, and C:
$$r_B(35) = A = 13; \quad r_B'(35) = B = 0; \quad r_B(60) = A + B \cdot 25 + C \cdot 25^2 = 27/4.$$

Thus $A = 13$, $B = 0$, and (from the last equation) $C = -1/100$.

3. (a) The appropriate integral is $\displaystyle\int_5^{55} r_B(t)\,dt = \int_5^{55} \left(13 - \frac{(t - 35)^2}{100}\right) dt = \frac{1600}{3}$. Thus Brown harvests about 533.33 bushels from $t = 5$ to $t = 55$.

(b) We want to estimate the integral $\int_5^{55} r_J(t)\,dt$ using T_5, the trapezoid rule with 5 subdivisions, each of length 10. Here's the result:
$$\left(\frac{r_J(5) + r_J(15)}{2} + \frac{r_J(15) + r_J(25)}{2} + \cdots + \frac{r_J(35) + r_J(45)}{2}\right) \cdot 10 = 525.$$

By this estimate, therefore, Jones harvests 525 bushels over the period.

5. (a) The linear function $f(x) = bx/a$ does the job. The length of its graph from $x = 0$ to $x = a$ is given by the integral
$$\int_0^a \sqrt{1 + f'(x)^2}\,dx = \int_0^a \sqrt{1 + b^2/a^2}\,dx = a\sqrt{1 + b^2/a^2} = \sqrt{a^2 + b^2}.$$

(An easier way to find the answer is to use the distance formula in the plane.)

(b) If $f(x) = x^2 + 1$, then $f'(x) = 2x$, so we want the integral $I = \displaystyle\int_0^1 \sqrt{1 + f'(x)^2}\,dx = \int_0^1 \sqrt{1 + 4x^2}\,dx$. We estimated this same integral in Example **??**, where we got $M_{20} \approx 1.479$.

(c) For each of the two curves in question we have $dy/dx = \cos x$, so the length integral is the same in each case:
$I = \displaystyle\int_0^\pi \sqrt{1 + f'(x)^2}\,dx = \int_0^\pi \sqrt{1 + \cos^2 x}\,dx$. For this integral, $M_{20} \approx 1.727$—that's a good estimate for the length of both curves.

(d) If $g(x) = f(x) + C$, for any constant C, then $g'(x) = f'(x)$, so the length of the g-graph from $x = a$ to $x = b$ is
$\displaystyle\int_a^b \sqrt{1 + g'(x)^2}\,dx = \int_a^b \sqrt{1 + f'(x)^2}\,dx$. The last quantity is independent of C.

Geometrically, the idea is that adding a constant C to f raises or lowers the graph of f, but *doesn't change its length*.

7. (a) $I = \displaystyle\int_0^1 \sqrt{1 + x}\,dx = \frac{2}{3}(1 + x)^{3/2}\Big]_0^1 = \frac{4\sqrt{2} - 2}{3} \approx 1.219.$

(b) I is the area under the curve $y = \sqrt{1 + x}$ from $x = 0$ to $x = 1$. (The region in question is more or less trapezoidal, with base 1 and altitudes 1 and $\sqrt{2}$.)

(c) We want a function f for which
$$I = \int_0^1 \sqrt{1 + x}\,dx = \int_0^1 \sqrt{1 + f'(x)^2}\,dx.$$

Let's look, therefore, for an f for which $f'(x)^2 = x$, or $f'(x) = \sqrt{x}$. *Any* antiderivative of $\sqrt{x}$ will do; let's use $f(x) = 2x^{3/2}/3$. Plotting this f over $[0, 1]$ gives a graph whose length appears to be around 1.2, as the previous part suggests.

(d) Any antiderivative of $\sqrt{x}$, i.e., any function of the form $f(x) = \frac{2}{3}x^{3/2} + C$, where C is a constant. (There are other possibilities, too. For instance, we could use $f(x) = -2x^{3/2}/3$, the *opposite* of the function in the previous part.

8.2 Finding volumes by integration

1. $V = \int_0^8 \pi \left(x^3\right)^2 dx = \dfrac{8^7\pi}{7}$

3. $V = \int_0^2 \pi \left((x+6)^2 - \left(x^3\right)^2\right) dx = \dfrac{1688\pi}{21}$

5. $V = \int_0^2 \pi \left(4^2 - \left(y^2\right)^2\right) dy = \dfrac{128\pi}{5}$

7. $V = \int_0^4 \pi \left(\sqrt{y}\right)^2 dy - \int_1^4 \pi \left(\log_2 y\right)^2 dy = \dfrac{16}{\ln 2} - 8 - \dfrac{6}{(\ln 2)^2} \approx 2.5949$

9. $V = \int_0^1 \pi \left(1^2 - (1 - \sqrt{x})^2\right) dx = \dfrac{5\pi}{6}$

11. $V = \int_{-4}^0 \pi \left(\left(2 + \sqrt{x+4}\right)^2 - \left(2 - \sqrt{x+4}\right)^2\right) dx + \int_0^5 \pi \left(\left(2 + \sqrt{x+4}\right)^2 - (2 - (2-x))^2\right) dx = \dfrac{128\pi}{3} + \dfrac{123\pi}{2} = $

13. At height y above the base, a cross section parallel to the base of the cone is a circle of radius $\frac{r}{h}(h - y)$. Thus, the volume of the cone is

$$V = \int_0^h \pi \left(\frac{r}{h}(h - y)\right)^2 dy = \frac{\pi r^2}{h^2} \int_{-h}^0 u^2 \, du = \frac{\pi r^2 h}{3}.$$

[The substitution $u = h - y$ was used to evaluate the integral.]

15. The area of an isosceles right triangle with hypotenuse h is $h^2/4$. Thus, the volume of the solid is $\int_1^4 \dfrac{dx}{4x^2} = \dfrac{3}{16}$.

17. $V = \pi \int_0^{\pi/4} \left(1 - \tan^2 y\right) dy = \pi \left(y - (\tan y - y)\right)\Big]_0^{\pi/4} = \pi \left(\dfrac{\pi}{2} - 1\right) \approx 1.7932.$

19. From the information given, the radius of the earth is $r = C/2\pi \approx 3{,}963$ miles.

(a) $V = \pi \int_{\sqrt{2}r/2}^r (r^2 - y^2) \, dy = \left(\frac{2}{3} - \frac{5\sqrt{2}}{12}\right) \pi r^3 \approx 1.1514 \times 10^{10}$ miles^3.

(b) $V = \pi \int_0^{\sqrt{2}r/2} (r^2 - y^2) \, dy = \frac{5\sqrt{2}\pi r^3}{12} \approx 1.1522 \times 10^{11}$ miles^3.

21. (a) $V = \pi \int_{-3}^{-1} \arctan^2 x \, dx$

(b) For any n, L_n overestimates V since $\left(\arctan^2 x\right)' = \dfrac{2\arctan x}{1 + x^2} < 0$ over the interval of integration (i.e., the integrand is a decreasing function).

23.

25.

27.

29.

8.3 Arclength

1. The integral is $\int_0^1 \sqrt{1+1}\,dx = \sqrt{2}$. This result is (of course) the same as that given by the usual distance formula.

3. (a) Here's a picture. Since the line connecting the endpoints of the curve C has length $\sqrt{(3-1)^2 + (109/12 - 7/12)^2} = \sqrt{305/4} \approx 8.7321$, the length of the curve is slightly more than this.

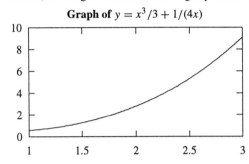

Graph of $y = x^3/3 + 1/(4x)$

(b) By the arclength formula, the length is the following integral. Watch the algebra very carefully:

$$\int_1^3 \sqrt{1 + (f')^2} = \int_1^3 \sqrt{1 + \left(x^2 - \frac{1}{4x^2}\right)^2}\,dx$$

$$= \int_1^3 \sqrt{1 + x^4 - \frac{1}{2} + \frac{1}{16x^4}}\,dx$$

$$= \int_1^3 \sqrt{x^4 + \frac{1}{2} + \frac{1}{16x^4}}\,dx$$

$$= \int_1^3 \sqrt{\left(x^2 + \frac{1}{4x^2}\right)^2}\,dx$$

$$= \int_1^3 \left(x^2 + \frac{1}{4x^2}\right)\,dx$$

$$= \left.\frac{x^3}{3} - \frac{1}{4x}\right]_1^3$$

$$= \frac{53}{6} \approx 8.833333333.$$

5. The shortest distance between two points A and B is a line. If the curve passing through the points A and B is not a line, then any polygonal path from A to B through intermediate points on the curve will be longer than the line segment connecting the two points.

7. length $= \int_0^1 \sqrt{1 + e^{2x}}\,dx = \frac{1}{2}\int_1^{e^2} \frac{\sqrt{1+u}}{u} = \frac{1}{2}\int_1^{e^2} \frac{1+u}{u\sqrt{1+u}} = \left.\left(\sqrt{1+u} + \frac{1}{2}\ln\left|\frac{\sqrt{1+u}-1}{\sqrt{1+u}-1}\right|\right)\right|_1^{e^2} \approx 2.003497110.$

8.4   Work

1. (a) work $= 115 \times 0.06$in-lbs $= 6.9$in-lbs $= 0.575$ft-lbs.

    (b) According to Hooke's Law, the force required to compress a spring x units is proportional to x. Since a force of 115 pounds compresses the spring by 0.06 inches, the constant of proportionality (the *spring constant* is $k = 115/0.06$. Therefore, a force of 175 pounds will compress the spring approximately 0.0913 inches and work $= 175 \times 0.0913$ in-lbs $= 15.978$ in-lbs $= 1$

3. Each parallel slice is a circular cylinder with cross-sectional area 25π and thickness Δx. Therefore,

$$\text{work} = \int_0^5 62.4 \cdot 25\pi \cdot (10 - x) \, dx = 58{,}500\pi \text{ ft-lbs} \approx 183{,}783 \text{ ft-lbs}.$$

5. For reasons as in Example 5, the necessary forces F_1 and F_2 for the two buckets are given, respectively, by

$$F_1(x) \quad = \quad 60 + 0.25 \cdot (60 - x) = 75 - \frac{x}{4};$$
$$F_2(x) \quad = \quad 50 + 0.25 \cdot (70 - x) = 68.75 - \frac{x}{4}.$$

    (The unit of force is pounds; x measures distance in feet from the bottom of the well.) To find the work in each case, we integrate:

$$W_1 \quad = \quad \int_0^{60} F_1(x) \, dx = \int_0^{60} \left(75 - \frac{x}{4} \right) dx = 4050 \text{ foot-pounds};$$
$$W_2 \quad = \quad \int_0^{70} F_2(x) \, dx = \int_0^{70} \left(68.75 - \frac{x}{4} \right) dx = 4200 \text{ foot-pounds}.$$

    Raising the *second* bucket takes a little more work.

7. Let x denote the number of inches of compression.

    (a) To find k, use the equation (implicit in the problem statement) $F(2) = 2k = 10$. It follows that $k = 5$ (and hence that $F(x) = 5x$.)

    (b) Compressing from 16 inches to 12 inches means compressing from $x = 2$ to $x = 6$. Thus the work done is

$$W = \int_2^6 F(x) \, dx = \int_2^6 5x \, dx = 80 \text{ inch-pounds}.$$

9. Let x denote the distance (in feet) that the spring is extended. Notice that since the chain weighs 20 pounds, we start with $x = 5$. (Draw a picture! Note, too, that other x-scales are possible.)

    The problem is to find $\int_5^7 F(x) \, dx$, where $F(x)$ is the net downward force necessary at a given x.

    Let's find a formula for $F(x)$; there are two main ingredients: (i) For any value of x, the spring exerts an *upward* force of $4x$ pounds. (ii) For a given value of x, the length of chain remaining above the floor is $10 - (x - 5) = 15 - x$ feet. (A diagram should make this convincing.) Since the chain weighs 2 pounds per foot, this length of chain exerts a *downward* force of $2(15 - x) = 30 - 2x$ pounds.

    Putting (i) and (ii) together means that the *net* downward force required for given x is $F(x) = 4x - (30 - 2x) = 6x - 30$ pounds. Thus the desired work is

$$W = \int_5^7 F(x) \, dx = \int_5^7 (6x - 30) \, dx = 12 \text{ foot-pounds}.$$

11. In each case, the work done is given by the integral $W = \int_a^b F(x) \, dx = \int_a^b k f'(x) \, dx$.

(a) If $k = 10$, $a = 0$, $b = 1$, $f(x) = x$, then $W = \displaystyle\int_0^1 10 \cdot dx = 10 \cdot x^3 \Big]_0^1 = 10$.

(b) If $k = 10$, $a = 0$, $b = 1$, $f(x) = x^3$, then $W = \displaystyle\int_0^1 10 \cdot 3x^2 \, dx = 10 \cdot x^3 \Big]_0^1 = 10$.

(c) If $k = 10$, $a = 0$, $b = 1$, $f(x) = x^n$, and n is *any* positive integer, then $W = \displaystyle\int_0^1 10 \cdot nx^{n-1} \, dx = 10 \cdot x^n \Big]_0^1 = 10$.

(d) We got the same answer each time. Mathematically, the point is that

$$W = \int_a^b F(x) \, dx = \int_a^b kf'(x) \, dx = k \ f(x)]_a^b = k(f(b) - f(a)).$$

Physically, the point is that in every case, the work is always the product of the object's weight and the *vertical* distance through which it travels. In a sense, then, the particular curve along which it travels makes no difference.

8.5   Present value

1. For any interest rate r, the present value of one $1 million payment, 23 years ahead, is $PV = 1,000,000 \cdot e^{-r \cdot 23}$.

   (a) If $r = 0.06$, then $PV = 1,000,000e^{-0.06 \cdot 23} \approx 251,579$ dollars.

   (b) If $r = 0.08$, then $PV = 1,000,000e^{-0.08 \cdot 23} \approx 155,817$ dollars.

   (c) To find the desired r we solve the equation $PV = 100,000 = 1,000,000e^{-r \cdot 23}$ for r. The result: $r = (\ln 10)/23 \approx 0.10011$—just a bit above 10%.

3. For any interest rate r (real or nominal) the present value of one $1 million payment, 23 years ahead, is $PV = 1,000,000 \cdot e^{-r \cdot 23}$.

   (a) If $r = 0.02$, then $PV = 1,000,000e^{-0.02 \cdot 23} \approx 631,284$ dollars.

   (b) If $r = 0.04$, then $PV = 1,000,000e^{-0.04 \cdot 23} \approx 398,519$ dollars.

   (c) To find Betty's r we solve the equation $PV = 200,000 = 1,000,000e^{-r \cdot 23}$ for r. The result: $r = (\ln 5)/23 \approx 0.07$. Thus Betty needs to find a *real* interest rate—after inflation—of 7%.

5. At any interest rate r, the present value formula for several future payments says, in this situation, that

$$PV = 40,000e^{-18r} + 42,000e^{-19r} + 44,000e^{-20r} + 46,000e^{-21r}.$$

   If $r = 0.06$, the formula (and some electronic help) give $PV \approx \$53,316.85$. If $r = 0.08$, $PV \approx \$36,119.66$.

7. (a) $\int_{10}^{20} 12,000e^{-0.06t} \, dt \approx \$49,523$

   (b) $\int_{10}^{20} 12,000e^{-0.08t} \, dt \approx \$37,115$

   (c) $\int_{10}^{20} 12,000 \, dt = \$120,000$

   (d) $\int_{10}^{20} 12,000e^{-0.04t} \, dt \approx \$66,297$

   (e) $\int_{10}^{20} 12,000e^{-0.06t} \, dt \approx \$49,523$

9. Using the midpoint rule with 50 subdivisions (and Maple): $PV_F \approx 21068.89$, $PV_T \approx 21203.64$, and $PV_I \approx 21067.78$.

9.1   Integration by parts

1. $\int xe^{2x}\,dx = \frac{1}{2}xe^{2x} - \frac{1}{4}e^{2x} + C$ $[du = dx, v = e^{2x}/2]$

3. $\int x\sec^2 x\,dx = x\tan(x) + \ln(\cos(x)) + C$ $[du = dx, v = \tan x]$

5. $\int x\sqrt{1+x}\,dx = \frac{2}{3}x(1+x)^{3/2} - \frac{4}{15}(1+x)^{5/2} + C$ $[du = dx, v = \frac{2}{3}(1+x)^{3/2}]$

7. Let $u = x$ and $dv = \cos^2 x\,dx = \frac{1}{2}(1 + \cos(2x))\,dx$. Then $du = dx$ and $v = \frac{1}{2}\left(x + \frac{1}{2}\sin(2x)\right)$. Therefore,
$$\int x\cos^2 x\,dx = \frac{1}{2}x^2 + \frac{1}{4}x\sin(2x) - \frac{1}{2}\int\left(x + \frac{\sin(2x)}{2}\right)dx = \frac{1}{4}x^2 + \frac{1}{4}x\sin(2x) + \frac{1}{8}\cos(2x) + C.$$

9.   (a) Let $u = \sin x$ and $dv = \sin x\,dx$. Then, $du = \cos x\,dx$ and $v = -\cos x$ so $\int \sin^2 x\,dx = -\sin x\cos x + \int\cos^2 x\,dx.$

  (b) $\int \sin^2 x\,dx = -\sin x\cos x + \int \cos^2 x\,dx = -\sin x\cos x + \int\left(1 - \sin^2 x\right)dx$. Therefore, $2\int \sin^2 x\,dx = x - \sin x\cos x$

11. $\int_0^1 xe^{-x}\,dx = -(1+x)e^{-x}\Big]_0^1 = 1 - 2e^{-1} \approx 0.26424.$ $[M_2 \approx 0.27449]$

13. $\int_{\pi/4}^{\pi/2} x\csc^2 x\,dx = -x\cot x + \ln(\sin x)\Big]_{\pi/4}^{\pi/2} = \pi/4 - \ln\left(\sqrt{2}/2\right) = \pi/4 + \ln\left(\sqrt{2}\right) \approx 1.13197.$ $[M_2 \approx 1.1188]$

15. $\int_1^4 e^{3x}\cos(2x)\,dx = \frac{3}{13}e^{3x}\cos(2x) + \frac{2}{13}e^{3x}\sin(2x)\Big]_1^4 \approx 19{,}307.$ $[M_2 \approx 24{,}861]$

17. $\int \arccos x\,dx = x\arccos x - \sqrt{1 - x^2} + C$ $[u = \arccos x, dv = dx]$

19. $\int (\ln x)^2\,dx = x(\ln x)^2 - 2x\ln x + 2x + C$ $[u = (\ln x)^2, dv = dx \text{ or } u = \ln x, dv = \ln x\,dx]$

21. $\int xe^x\sin x\,dx = \frac{1}{2}(1 - x)e^x\cos x + \frac{1}{2}xe^x\sin x + C.$ $[u = x, dv = e^x\sin x\,dx]$

23. $\int x\arctan x\,dx = \frac{1}{2}\left(x^2\arctan x - x + \arctan x\right) + C$
Use integration by parts with $u = \arctan x$ and $dv = x\,dx$ to show that $\int x\arctan x\,dx = \frac{1}{2}\left(x^2\arctan x - \int x^2/(1 + x^2)\,dx\right)$. The last antiderivative can be found using the algebraic identity
$$\frac{x^2}{1 + x^2} = \frac{(1 + x^2) - 1}{1 + x^2} = 1 - \frac{1}{1 + x^2}.$$

25. $\int x^5\sin\left(x^3\right)\,dx = \frac{1}{3}\sin\left(x^3\right) - \frac{1}{3}x^3\cos\left(x^3\right) + C$
First substitute $w = x^3$, $dw = 3x^2\,dx$, to get the new integral $\frac{1}{3}\int w\sin w\,dw$. Now use parts, with $u = w, dv = \sin w\,dw$.

27. $\int \sqrt{x}\,e^{-\sqrt{x}}\,dx = -2e^{-\sqrt{x}}\left(x + 2\sqrt{x} + 2\right) + C$
First substitute $w = \sqrt{x}$, $w^2 = x$, $2w\,dw = dx$. This gives the new integral $2\int w^2 e^{-w}\,dw$. Finding the latter integral requires using integrations by parts twice.

29. $\displaystyle\int \frac{\arctan\left(\sqrt{x}\right)}{\sqrt{x}} = 2\sqrt{x}\arctan\left(\sqrt{x}\right) - \ln(1+x) + C$

Substitute $w = \sqrt{x}$, then use integration by parts with $u = \arctan w$ and $dv = dw$.

31. (a) $I_1 = \displaystyle\int x(\ln x)^1\, dx$. Let $u = \ln x$, $dv = x\, dx$; $du = \dfrac{1}{x}\, dx$; $v = \dfrac{x^2}{2}$. Then

$$I_1 = \int x(\ln x)^1\, dx = \frac{x^2}{2}\ln x - \int \frac{x}{2}\, dx = \frac{x^2}{2}\ln x - \frac{x^2}{4} + C$$

(b) $I_n = \displaystyle\int x(\ln x)^n\, dx$. If we let $u = (\ln x)^n$; $dv = x\, dx$; $du = n(\ln x)^{n-1} \cdot \dfrac{1}{x}\, dx$; $v = \dfrac{x^2}{2}$. then we get

$$I_n = \frac{(\ln x)^n x^2}{2} - \frac{n}{2}\int x(\ln x)^{n-1}\, dx = \frac{x^2}{2}(\ln x)^n - \frac{n}{2}I_{n-1}$$

(c) By reduction formula plus the fact that $I_1 = \dfrac{x^2}{2}\ln x - \dfrac{x^2}{4} + C$, we have

$$I_2 = \frac{x^2}{2}(\ln x)^2 - I_1 = \frac{x^2}{2}(\ln x)^2 - \left(\frac{x^2}{2}\ln x - \frac{x^2}{4}\right) + C = \frac{x^2}{2}\left((\ln x)^2 - \ln x + \frac{1}{2}\right) + C.$$

Similarly,

$$\begin{aligned} I_3 &= \frac{x^2}{2}(\ln x)^3 - \frac{3}{2}I_2 \\ &= \frac{x^2}{2}\left((\ln x)^3 - \frac{3}{2}(\ln x)^2 + \frac{3}{2}\ln x - \frac{3}{4}\right) + C \end{aligned}$$

(d) This is an immediate consequence of the Fundamental Theorem of Calculus.

(e) Carrying out the differentiation on the right side of the identity

$$\begin{aligned} \text{(Right Side)}' &= x \cdot (\ln x)^n + \frac{x^2}{2} \cdot n(\ln x)^{n-1} \cdot \frac{1}{x} - \frac{n}{2}x(\ln x)^{n-1} \\ &= x \cdot (\ln x)^n \quad \text{(after a bit of algebra.)} \end{aligned}$$

33. Let $u = (\ln x)^n$ and $dv = dx$.

35. (a) $-\dfrac{1}{2}\cos\left(x^2\right) + C$

(b) $-\dfrac{1}{2}x^2\cos\left(x^2\right) + \dfrac{1}{2}\sin\left(x^2\right) + C$

(c) $\dfrac{1}{2}x^2\sin\left(x^2\right) + \dfrac{1}{2}\cos\left(x^2\right) + C$

(d) $\displaystyle\int x^2\cos\left(x^2\right)\, dx = \dfrac{1}{2}x\sin\left(x^2\right) - \dfrac{1}{2}\int \sin\left(x^2\right)$. Since the expression on the left side of the equals sign is not elementary, neither is the expression on the right side.

37. An integration by parts with $u = f(x)$ and $dv = \sin x\, dx$ shows that

$$\int_0^\pi f(x)\sin x\, dx = -f(x)\cos x \Big]_0^\pi + \int_0^\pi f'(x)\cos x\, dx = f(\pi) + f(0) + \int_0^\pi f'(x)\cos x\, dx.$$

Now an integration by parts with $u = f'(x)$ and $dv = \cos x\, dx$ shows that

$$\int_0^\pi f'(x)\cos x\, dx = f'(x)\sin x \Big]_0^\pi - \int_0^\pi f''(x)\sin x\, dx = -\int_0^\pi f''(x)\sin x\, dx.$$

Combining these results, we have

$$6 = \int_0^\pi f(x) \sin x \, dx + \int_0^\pi f''(x) \sin x \, dx = f(\pi) + f(0) = f(\pi) + 2.$$

From this it follows that $f(\pi) = 4$.

9.2 Partial fractions

1. $\displaystyle \int \frac{5x+7}{(x+1)(x+2)}\,dx = 2\ln|x+1| + 3\ln|x+2| + C$

3. $\displaystyle \int \frac{5x+7}{(x+1)(x+2)}\,dx = 2\ln|x+1| + 3\ln|x+2| + C$

5. Let $I = \displaystyle\int \frac{x^2-1}{x(x^2+4)}\,dx$. First write $\dfrac{x^2-1}{x(x^2+4)} = \dfrac{a}{x} + \dfrac{bx+c}{x^2+4}$. Solving for $a, b,$ and c gives $a = -\frac{1}{4}, b = \frac{5}{4}, c = 0$. Hence

$$I = -\frac{1}{4}\ln|x| + \frac{5}{8}\ln\left|x^2+4\right| + C$$

7. First complete the square: $I = \displaystyle\int \frac{x}{x^2+2x+6}\,dx = \int \frac{x}{(x+1)^2+5}\,dx$. Now let $u = x+1, x = u-1, dx = du$. Then,

$$
\begin{aligned}
I &= \int \frac{u-1}{u^2+5}\,du \\
&= \int \frac{u}{u^2+5}\,du - \int \frac{1\,du}{u^2+5} \\
&= \frac{1}{2}\ln\left|u^2+5\right| - \frac{1}{\sqrt{5}}\tan^{-1}\left(\frac{u}{\sqrt{5}}\right) + C \\
&= \frac{1}{2}\ln\left|x^2+2x+6\right| - \frac{1}{\sqrt{5}}\tan^{-1}\left(\frac{x+1}{\sqrt{5}}\right) + C
\end{aligned}
$$

9. $\displaystyle \int \frac{x^4}{x^4-1}\,dx = \int \left(1 + \frac{1}{4}\cdot\frac{1}{x-1} - \frac{1}{4}\cdot\frac{1}{x+1} - \frac{1}{2}\cdot\frac{1}{1+x^2}\right)dx =$
$x + \frac{1}{4}\ln(x-1) - \frac{1}{4}\ln(x+1) - \frac{1}{2}\arctan x + C$

11. $\displaystyle \int \frac{2x+1}{(x-2)(x+3)}\,dx = \int \left(\frac{1}{x-2} + \frac{1}{x+3}\right)dx = \ln(x-2) + \ln(x+3) + C$

13. $\displaystyle \int \frac{x^2+x}{(x^2+4)^2}\,dx = \int \left(\frac{1}{x^2+4} + \frac{x-4}{(x^2+4)^2}\right)dx = \frac{1}{4}\arctan(x/2) - \frac{1}{2}\cdot\frac{x+1}{x^2+4} + C$

15. $\displaystyle \int \frac{x^3}{x^2-1}\,dx = \int \left(x + \frac{1}{2}\cdot\frac{1}{x-1} + \frac{1}{2}\cdot\frac{1}{x+1}\right)dx = \frac{1}{2}x^2 + \frac{1}{2}\ln(x-1) + \frac{1}{2}\ln(x+1) + C$

17. $\displaystyle \int \frac{x^2}{(x^2+1)(x+1)^2}\,dx = \int \left(\frac{1}{2}\cdot\frac{1}{(1+x)^2} - \frac{1}{2}\cdot\frac{1}{x+1} + \frac{1}{2}\cdot\frac{x}{x^2+1}\right)dx =$
$-\frac{1}{2}\cdot\frac{1}{x+1} - \frac{1}{2}\ln(x+1) + \frac{1}{4}\ln(x^2+1) + C$

9.3 Trigonometric antiderivatives

1. No—the two answers are equal. To see this, use the identity $\cos^2 x = 1 - \sin^2 x$.

3. $\displaystyle \int \cos^4 x \sin^2 x \, dx = \int \left(\cos^4 x - \cos^6 x \right) dx = \frac{1}{16}x + \frac{1}{16}\cos x \sin x + \frac{1}{24}\cos^3 x \sin x - \frac{1}{6}\sin x \cos^5 x + C$

5. $\displaystyle \int \sin(2x) \cos^2 x \, dx = \frac{1}{2} \int \sin(2x)\left(1 + \cos(2x)\right) dx = -\frac{1}{4}\cos(2x) - \frac{1}{8}\cos^2(2x) + C$

7. $\displaystyle \int \tan^4 x \, dx = \frac{1}{3}\tan^3 x - \tan x + x + C$

9. $\displaystyle \int \sec^3 x \tan^2 x \, dx = \int \left(\sec^5 x - \sec^3 x \right) dx = \frac{1}{4}\sec^3 x \tan x - \frac{1}{8}\sec x \tan x - \frac{1}{8}\ln(\sec x + \tan x) + C$

11. $\displaystyle \int \sqrt{\cos x}\, \sin^5 x \, dx = \int \sqrt{\cos x}\left(1 - \cos^2 x\right)^2 \sin x \, dx = -\frac{2}{3}(\cos x)^{3/2} + \frac{4}{7}(\cos x)^{7/2} - \frac{2}{11}(\cos x)^{11/2} + C$

13. $\displaystyle \int \frac{dx}{\left(x^2 + 4\right)^2} \quad \rightarrow \quad \frac{1}{8}\int \frac{\sec^2 t}{\left(1 + \tan^2 t\right)^2}\, dt$

$\displaystyle \qquad = \quad \frac{1}{8}\int \cos^2 t \, dt$

$\displaystyle \qquad = \quad \frac{t}{16} + \frac{\sin(2t)}{32} + C = \frac{t}{16} + \frac{\sin t \cos t}{16} + C$

$\displaystyle \qquad \rightarrow \quad \frac{1}{16}\arctan(x/2) + \frac{x}{8\left(4 + x^2\right)}$

$[x = \tan t, dx = \sec^2 t \, dt, \sin t = x/\sqrt{4 + x^2}, \cos t = 2/\sqrt{4 + x^2}]$

15. $\displaystyle \int \frac{x^2}{\sqrt{9 - x^2}}\, dx \quad \rightarrow \quad 9\int \sin^2 t \, dt$

$\displaystyle \qquad = \quad \frac{9t}{2} - \frac{9}{4}\sin(2t) + C$

$\displaystyle \qquad = \quad \frac{9t}{2} - \frac{9}{2}\sin t \cos t + C$

$\displaystyle \qquad \rightarrow \quad \frac{9}{2}\arcsin(x/3) - \frac{1}{2}x\sqrt{9 - x^2} + C$

$[x = 3\sin t, dx = 3\cos t \, dt, \cos t = \sqrt{1 - x^2/9}]$

17. $\displaystyle \int \frac{dx}{x^2\sqrt{x^2 - 4}} \quad \rightarrow \quad \frac{1}{4}\int \cos t \, dt$

$\displaystyle \qquad = \quad \frac{1}{4}\sin t + C$

$\displaystyle \qquad \rightarrow \quad \frac{\sqrt{x^2 - 4}}{4x}$

$[x = 2\sec t, dx = 2\sec t \tan t \, dt, \sin t = \sqrt{1 - 4/x^2}]$

19. $\displaystyle \int \frac{dx}{x^2\sqrt{x^2 + 1}} \quad \rightarrow \quad \int \frac{\cos t}{\sin^2 t}\, dt$

$\displaystyle \qquad = \quad -\frac{1}{\sin t} + C$

$\displaystyle \qquad \rightarrow \quad -\frac{\sqrt{x^2 + 1}}{x} + C$

$[x = \tan t, dx = \sec^2 t \, dt, \sin t = x/\sqrt{x^2 + 1}]$

21. First, note that $\int \dfrac{x+2}{x\left(x^2+1\right)} \, dx = \int \dfrac{dx}{x^2+1} + 2 \int \dfrac{dx}{x\left(x^2+1\right)} = \arctan x + 2 \int \dfrac{dx}{x\left(x^2+1\right)}.$

$$
\begin{aligned}
\int \frac{dx}{x\left(x^2+1\right)} \quad &\rightarrow \quad \int \frac{\cos t}{\sin t} \, dt \\
&= \quad \ln|\sin t| + C \\
&\rightarrow \quad \ln\left|\frac{x}{\sqrt{1+x^2}}\right| + C
\end{aligned}
$$

$[x = \tan t, dx = \sec^2 t, \sin t = x/\sqrt{1+x^2}]$

Therefore, $\int \dfrac{x+2}{x\left(x^2+1\right)} \, dx = \arctan x + 2\ln\left|\dfrac{x}{\sqrt{1+x^2}}\right| + C.$

23. (a) Let $x = \sec t$. When $x > 0$, $\sqrt{x^2-1} = \tan t$ (since $\tan t > 0$). Thus,

$$
\begin{aligned}
\int_1^2 \frac{\sqrt{x^2-1}}{x} \, dx \quad &= \quad \int_0^{\pi/3} \tan^2 t \, dt \\
&= \quad \tan t - t \, \Big]_0^{\pi/3} \\
&= \quad \sqrt{3} - \pi/3
\end{aligned}
$$

(b) Let $x = \sec t$. When $x < 0$, $\sqrt{x^2-1} = -\tan t$ (since $\tan t < 0$). Thus,

$$
\begin{aligned}
\int_{-2}^{-1} \frac{\sqrt{x^2-1}}{x} \, dx \quad &= \quad -\int_{2\pi/3}^{\pi} \tan^2 t \, dt \\
&= \quad t - \tan t \, \Big]_{2\pi/3}^{\pi} \\
&= \quad \pi/3 - \sqrt{3}
\end{aligned}
$$

10.1   When is an integral improper?

1. (a) The interval of integration is infinite.

    (b) The integrand is unbounded as $x \to 1^-$.

    (c) The integrand is unbounded as $x \to 1^+$.

    (d) The integrand is unbounded near $x = 2$.

    (e) The integrand is unbounded as $x \to \frac{\pi}{2}^-$

    (f) The integrand is unbounded near $x = \pi$.

3. $\displaystyle\int_0^\infty \frac{dx}{x^2} = \int_0^1 \frac{dx}{x^2} + \int_1^\infty \frac{dx}{x^2}$. Since $\displaystyle\int_0^1 \frac{dx}{x^2} = \infty$, the original improper integral diverges.

5. $\displaystyle\int_e^\infty \frac{dx}{x\,(\ln x)^2} = \lim_{t\to\infty} \frac{-1}{\ln x}\bigg]_e^t = \lim_{t\to\infty}\left(1 - \frac{1}{\ln t}\right) = 1$

7. $\displaystyle\int_0^4 \frac{dx}{\sqrt{x}} = \lim_{t\to 0^+}\int_t^4 \frac{dx}{\sqrt{x}} = \lim_{t\to 0^+} 2\sqrt{x}\bigg]_t^4 = \lim_{t\to 0^+}\left(4 - 2\sqrt{t}\right) = 4$

9. $\displaystyle\int_\pi^\infty e^{-x}\sin x\,dx = \lim_{t\to\infty} -\tfrac{1}{2}e^{-x}(\sin x + \cos x)\bigg]_\pi^t = -\tfrac{1}{2}e^{-\pi}$

11. $\displaystyle I = \int_0^\infty f(x)\,dx = \int_0^a f(x)\,dx + \int_a^\infty f(x)\,dx \implies \left|I - \int_0^a f(x)\,dx\right| = \left|\int_a^\infty f(x)\,dx\right| \le 0.0001.$

13. $\displaystyle\int_a^\infty \frac{dx}{x^2+1} = \frac{\pi}{2} - \arctan a \le 10^{-5}$ when $a \ge 100,000$.

15. (a) $\displaystyle\lim_{a\to\infty}\int_{-a}^a x\,dx = \lim_{a\to\infty} \tfrac{1}{2}x^2\bigg]_{-a}^a = \lim_{a\to\infty}\left(\tfrac{1}{2}(a)^2 - \tfrac{1}{2}(-a)^2\right) = \lim_{a\to\infty}(0) = 0$

    (b) $\displaystyle\int_{-\infty}^\infty x\,dx = \lim_{s\to-\infty}\int_s^0 x\,dx + \lim_{t\to\infty}\int_0^t x\,dx$. Since neither $\displaystyle\lim_{s\to-\infty}\int_s^0 x\,dx$ nor $\displaystyle\lim_{t\to\infty}\int_0^t x\,dx$ exists, the original improper integral diverges.

17. Converges. $\displaystyle\int_0^\infty \frac{\arctan x}{1+x^2}\,dx = \lim_{t\to\infty} \frac{1}{2}(\arctan t)^2 = \pi^2/8$

19. Converges. $\displaystyle\int_3^\infty \frac{x}{(x^2-4)^3}\,dx = \lim_{t\to\infty} \frac{1}{4}\left(\frac{1}{25} - \frac{1}{(t^2-4)^2}\right) = \frac{1}{100}$

21. Converges. $\displaystyle\int_0^8 \frac{dx}{\sqrt[3]{x}} = \lim_{t\to 0^+} \frac{3}{2}\left(8^{2/3} - t^{2/3}\right) = 6$

23. Converges. $\displaystyle\int_2^3 \frac{x}{\sqrt{3-x}}\,dx = \lim_{t\to 3^-}\left(\frac{2}{3}(3-t)^{3/2} - 6\sqrt{3-t} + \frac{16}{3}\right) = \frac{16}{3}$

25. Diverges. $\displaystyle\int_1^\infty \frac{dx}{x(\ln x)^2} = \int_1^2 \frac{dx}{x(\ln x)^2} + \int_2^\infty \frac{dx}{x(\ln x)^2} = \lim_{s\to 1^+}\left(\frac{1}{\ln s} - \frac{1}{\ln 2}\right) + \lim_{t\to\infty}\left(\frac{1}{\ln 2} - \frac{1}{\ln t}\right) = \infty$

27. Diverges. $\displaystyle\int_0^\infty \frac{dx}{e^x - 1} = \int_0^1 \frac{dx}{e^x - 1} + \int_1^\infty \frac{dx}{e^x - 1}$ and $\displaystyle\int_0^1 \frac{dx}{e^x - 1} = \infty$

29. Converges. $\displaystyle\int_{-\infty}^\infty \frac{dx}{e^x + e^{-x}} = \int_{-\infty}^\infty \frac{e^x}{e^{2x}+1}\,dx = \arctan\left(e^x\right)\bigg]_{-\infty}^\infty = \frac{\pi}{2}$

31. Converges. $\displaystyle\int_0^1 \frac{e^{-\sqrt{x}}}{\sqrt{x}}\,dx = -2e^{-\sqrt{x}}\Big]_0^1 = 2 - 2/e$

33. (a) $\displaystyle\int_1^\infty \frac{dx}{x} = \lim_{t\to\infty}\int_1^t \frac{dx}{x} = \lim_{t\to\infty}\ln t = \infty$

   (b) If $p > 1$, $\displaystyle\int_1^\infty \frac{dx}{x^p} = \lim_{t\to\infty}\frac{1-t^{1-p}}{p-1} = \frac{1}{p-1}$ since $\lim_t \to \infty t^{1-p} = 0$.

   (c) If $p < 1$, $\displaystyle\int_1^\infty \frac{dx}{x^p} = \lim_{t\to\infty}\frac{1-t^{1-p}}{p-1} = \infty$ since $\lim_t \to \infty t^{1-p} = \infty$.

35. $\displaystyle\int_0^1 \frac{dx}{x^p} = \lim_{t\to 0^+}\int_t^1 \frac{dx}{x^p} = \lim_{t\to 0^+}\frac{1-t^{1-p}}{p-1}$. This limit is a finite number only when $p < 1$. Thus, $\displaystyle\int_0^1 \frac{dx}{x^p}$ converges when $p < 1$ and diverges when $p \geq 1$.

37. $\displaystyle\int\left(\frac{2x}{x^2+1} - \frac{C}{2x+1}\right)dx = \ln(x^2+1) - \frac{C}{2}\ln(2x+1) = \ln\left(\frac{x^2+1}{\sqrt{(2x+1)^C}}\right)$. Thus, the improper integral converges to $-2\ln 2$ when $C = 4$ and diverges for all other values of C.

39. $\displaystyle\int\left(\frac{Cx^2}{x^3+1} - \frac{1}{3x+1}\right)dx = \frac{C}{3}\ln(x^3+1) - \frac{1}{3}\ln(3x+1)$. Thus, the improper integral converges to $-\frac{1}{3}\ln 3 + \frac{5}{9}\ln 2$ when $C = \frac{1}{3}$ and diverges for all other values of C.

41. $\displaystyle\int\left(\frac{x}{x^2+1} - \frac{C}{3x+1}\right)dx = \frac{1}{2}\ln\left(x^2+1\right) - \frac{C}{3}\ln(3x+1)$. Thus, the improper integral converges to $-\ln 3$ when $C = 3$ and diverges for all other values of C.

43. $\displaystyle\int_1^\infty \frac{x}{x^3+1}\,dx = \lim_{t\to\infty}\int_1^t \frac{x}{x^3+1}\,dx = \lim_{t\to\infty} -\int_1^{1/t} \frac{du}{1+u^3} = \int_0^1 \frac{du}{1+u^3}$

45. Use the substitution $u = 1/x$:

$$
\begin{aligned}
\int_0^\infty \frac{dx}{1+x^4} &= \int_0^1 \frac{dx}{1+x^4} + \int_1^\infty \frac{dx}{1+x^4}\\
&= \lim_{s\to 0^+}\int_s^1 \frac{dx}{1+x^4} + \lim_{t\to\infty}\int_1^t \frac{dx}{1+x^4}\\
&= \lim_{s\to 0^+} -\int_{1/s}^1 \frac{u^2}{u^4+1}\,du + \lim_{t\to\infty} -\int_1^{1/t} \frac{u^2}{u^4+1}\,du\\
&= \int_1^\infty \frac{u^2}{u^4+1}\,du + \int_0^1 \frac{u^2}{u^4+1}\,du\\
&= \int_0^\infty \frac{u^2}{u^4+1}\,du
\end{aligned}
$$

47. $\displaystyle\int_0^\infty \frac{x\ln x}{1+x^4}\,dx = 0$.

   Using the substitution $u = 1/x$: $\displaystyle\int_1^\infty \frac{x\ln x}{1+x^4}\,dx = -\int_0^1 \frac{u\ln u}{u^4+1}\,dx$.

10.2 Detecting convergence, estimating limits

1. (a) For every $x \in \mathbb{R}$, $-1 \le \sin x \le 1$

   (b) $\displaystyle\int_2^\infty \frac{dx}{x + \sin x} \ge \int_2^\infty \frac{dx}{x - 1} = \infty$

3. (a) $0 \le \sqrt{x} \le x^2/2$ for all $x \ge 2$, so $x^2/2 = x^2 - x^2/2 \le x^2 - \sqrt{x} \le x^2$ for all $x \ge 2$.

   (b) $\displaystyle 0 \le \int_3^\infty \frac{dx}{x^2 - \sqrt{x}} \le 2 \int_3^\infty \frac{dx}{x^2} = \frac{1}{3}$

5. (a) Note that $\displaystyle\frac{x}{\sqrt{1+x^3}} = \frac{1}{\sqrt{x^{-2}+x}}$. Therefore, since $0 < x^{-2} \le x$ when $x \ge 1$, $\displaystyle\frac{1}{\sqrt{2x}} \le \frac{x}{\sqrt{1+x^3}} \le \frac{1}{\sqrt{x}}$.

   (b) The improper integral **diverges** since

   $$0 \le \int_0^\infty \frac{x}{\sqrt{1+x^3}}\,dx = \int_0^1 \frac{x}{\sqrt{1+x^3}}\,dx + \int_1^\infty \frac{x}{\sqrt{1+x^3}}\,dx$$
   $$\ge \int_0^1 \frac{x}{\sqrt{1+x^3}}\,dx + \int_1^\infty \frac{dx}{\sqrt{x}}$$

   and $\displaystyle\int_1^\infty \frac{dx}{\sqrt{x}}$ diverges.

7. (a) When $x \ge 1$, $\sqrt{x} \ge 1$. Therefore, $x^4 \le \sqrt{x} \cdot x^4 = \sqrt{x} \cdot \sqrt{x^8} = \sqrt{x^9} < \sqrt{1+x^9}$ when $x \ge 1$.
   When $x \ge 1$, $x^9 \ge 1$. Therefore, $\sqrt{1+x^9} \le \sqrt{x^9+x^9} = \sqrt{2}x^{9/2}$.

   (b) Converges. $\displaystyle 0 \le \int_0^\infty \frac{dx}{\sqrt{1+x^9}} = \int_0^1 \frac{dx}{\sqrt{1+x^9}} + \int_1^\infty \frac{dx}{x^4} < \infty$

9. (a) Let $f(x) = \sin x - x/2$. Then $f(0) = 0$ and $f'(x) = \cos x - 1/2 > 0$ when $0 \le x \le 1$. Therefore, $f(x) > 0$ when $0 \le x \le 1$.

   (b) $\displaystyle 0 \le \int_0^1 \frac{dx}{\sqrt{\sin x}} \le \sqrt{2} \int_0^1 \frac{dx}{\sqrt{x}} = 2\sqrt{2}$.

11. $\displaystyle 0 < \int_a^\infty \frac{dx}{x^4\sqrt{2x^3+1}} < \frac{1}{\sqrt{2}} \int_a^\infty \frac{dx}{x^{11/2}} = \frac{\sqrt{2}}{9} a^{-9/2} \le 10^{-5}$ when $a \ge \left(\frac{2\times 10^{10}}{81}\right)^{1/9} \approx 8.5606$.

    Thus, $\displaystyle\int_1^9 \frac{dx}{x^4\sqrt{2x^3+1}}$ approximates $\displaystyle\int_1^\infty \frac{dx}{x^4\sqrt{2x^3+1}}$ within 10^{-5}.

13. $\displaystyle 0 < \int_a^\infty \frac{e^{-x}}{2+\cos x}\,dx \le \int_a^\infty e^{-x}\,dx = e^{-a} \le 10^{-5}$ when $a \ge \ln(10^5) \approx 11.513$. Thus, $\displaystyle\int_0^{12} \frac{e^{-x}}{2+\cos x}\,dx$ approximates $\displaystyle\int_0^\infty \frac{e^{-x}}{2+\cos x}\,dx$ within 10^{-5}.

15. $\displaystyle\left|\int_a^\infty e^{-x^2}\sin x\,dx\right| \le \int_a^\infty \left|e^{-x^2}\sin x\right|\,dx \le \int_a^\infty e^{-x^2}\,dx \le \int_a^\infty xe^{-x^2}\,dx = \tfrac{1}{2}e^{-a^2} \le 0.0025$

    when $a \ge \sqrt{\ln 200} \approx 2.3018$. Thus, $\displaystyle\int_0^3 e^{-x^2}\sin x\,dx$ approximates $\displaystyle\int_0^\infty e^{-x^2}\sin x\,dx$ with an error no greater than 0.0025.

    Using $K_2 = 2.25$ (or $K_4 = 21$), we find that a midpoint rule estimate computed with $n \ge 32$ (or a Simpson's rule estimate computed with $n \ge 12$) approximates $\displaystyle\int_0^3 e^{-x^2}\sin x\,dx$ with an error no greater than 0.0025. Therefore, $M_{32} \approx 0.42480$ and $S_{12} \approx 0.42460$ are estimates of $\displaystyle\int_0^\infty e^{-x^2}\sin x\,dx$ guaranteed to be accurate within 0.005.

17. $\displaystyle\int_a^\infty \frac{\arctan x}{\left(1+x^2\right)^4}\,dx \le \frac{\pi}{2}\int_a^\infty x^{-8}\,dx = \frac{\pi}{14a^7} \le 0.0025$ when $a \ge 1.9011$. Thus, $\displaystyle\int_1^2 \frac{\arctan x}{\left(1+x^2\right)^4}\,dx$ approximates $\displaystyle\int_1^\infty \frac{\arctan x}{\left(1+x^2\right)}$

with an error no greater than 0.0025.

Using $K_2 = 0.6$ and $K_4 = 2.25$, we find that the midpoint rule with $n \ge 4$ and Simpson's rule with $n \ge 2$ approximate $\displaystyle\int_1^2 \frac{\arctan x}{\left(1+x^2\right)^4}\,dx$ with an error no greater than 0.0025. Therefore, $M_4 \approx 0.013483$ and $S_2 \approx 0.014349$ are estimates of $\displaystyle\int_1^\infty \frac{\arctan x}{\left(1+x^2\right)^4}\,dx$ guaranteed to be accurate with 0.005.

19. **Diverges.**   Since $x^4 \le x$ when $0 \le x \le 1$, $\displaystyle\int_0^\infty \frac{dx}{x^4+x} = \int_0^1 \frac{dx}{x^4+x} + \int_1^\infty \frac{dx}{x^4+x} > \int_0^1 \frac{dx}{2x} + \int_0^1 \frac{dx}{x^4+x}$. Since $\displaystyle\int_0^1 \frac{dx}{2x}$ diverges, the original improper integral also diverges.

21. **Diverges** because $\displaystyle\lim_{x\to\infty} e^{\sin x}$ does not exist ($e^{\sin x} \ge e^{-1}$ for all x).

23. **Converges.** $0 \le \displaystyle\int_0^\infty \frac{dx}{x+e^x} \le \int_0^\infty e^{-x}\,dx = 1.$

25. **Converges.** $0 \le \displaystyle\int_1^\infty \frac{dx}{\sqrt[3]{x^6+x}} < \int_1^\infty \frac{dx}{x^2} = 1.$

27. **Converges.** $\left|\displaystyle\int_2^\infty \frac{\sin x}{x^2\sqrt{x-1}}\,dx\right| \le \int_2^\infty \frac{dx}{x^2} = 1/2.$

29. **Diverges.**   Since $x > \ln x$ for all $x \ge 1$, $\displaystyle\int_3^\infty \frac{x}{\ln x}\,dx > \int_3^\infty \frac{x}{x}\,dx = \int_3^\infty dx = \infty.$

31. **Converges.** $0 \le \displaystyle\int_0^\infty \frac{dx}{\sqrt{x+x^3}} = \int_0^1 \frac{dx}{\sqrt{x+x^3}} + \int_1^\infty \frac{dx}{\sqrt{x+x^3}} \le \int_0^1 \frac{dx}{\sqrt{x}} + \int_1^\infty \frac{dx}{\sqrt{x^3}} = 2 + 2 = 4$

33. (a) I is an improper integral because the interval of integration is unbounded (i.e., infinitely long).

   (b) Since $I = \displaystyle\int_0^\infty f(x)\,dx = \int_0^1 f(x)\,dx + \int_1^\infty f(x)\,dx$, I converges if and only if $\displaystyle\int_1^\infty f(x)\,dx$ converges. The comparison test can be used to show that this integral converges: since $0 \le e^{-x^3}\cos^2 x \le e^{-x^3} \le x^2 e^{-x^3}$ when $x \ge 1$, it follows that $0 \le \displaystyle\int_1^\infty f(x)\,dx \le \int_1^\infty x^2 e^{-x^3}\,dx = \frac{1}{3e}$ and, therefore, $\displaystyle\int_1^\infty f(x)\,dx$ converges.

   (c) $I = \displaystyle\int_0^a f(x)\,dx + \int_a^\infty f(x)\,dx \le \int_0^a f(x)\,dx + \int_a^\infty x^2 e^{-x^3}\,dx = \int_0^a f(x)\,dx + \frac{1}{3}e^{-a^3}$ for any $a \ge 1$. Now, $\frac{1}{3}e^{-a^3} \le 0.0025$ when $a \ge \sqrt[3]{-\ln 0.0075} \approx 1.6977$ so $\displaystyle\int_0^2 f(x)\,dx$ approximates I within 0.0025. Thus, an estimate of $\displaystyle\int_0^2 f(x)\,dx$ that is in error by no more than 0.0025 will be an estimate of I that is guaranteed to be correct within $0.0025 + 0.0025 = 0.005$.

   Using the derivative bounds given, it can be shown that $\displaystyle\int_0^2 f(x)\,dx$ is approximated within 0.0025 by M_n when $n \ge 20$ and by S_n when $n \ge 8$. $M_{20} \approx 0.64425$; $S_8 \approx 0.64441$. Thus, $|I - 0.64425| \le 0.005$.

10.3   Improper integrals and probability

1. Graphs appear below; for reference, the the standard normal graph appears, too:

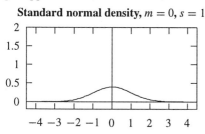

Standard normal density, $m = 0$, $s = 1$

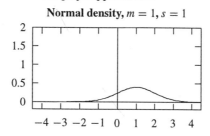

Normal density, $m = 1$, $s = 1$

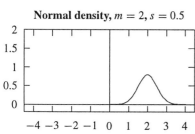

Normal density, $m = 2$, $s = 0.5$

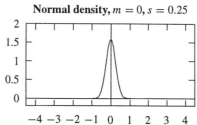

Normal density, $m = 0$, $s = 0.25$

Notice the similarities among all the graphs—m locates the center; s determines the "spread."

3. We want $I_1 = \displaystyle\int_{6.5}^{8.5} f(x)\,dx$ and $I_2 = \displaystyle\int_{10}^{\infty} f(x)\,dx = 0.5 - \int_{7.5}^{10} f(x)\,dx$, where (given that $m = 7.5$, $s = 1$)

$$f(x) = \frac{1}{\sqrt{2\pi}} \exp\left(-\frac{(x - 7.5)^2}{2}\right).$$

Using the midpoint rule, M_{20} gave (approximate) answers of 0.6829 and 0.0062, respectively. Applying the trapezoid rule with 20 subdivisions to I_1 and I_2 respectively gives $I_1 \approx T_{20} \approx 0.6823$; $I_2 \approx T_{20} \approx 0.0063$. These numerical results are consistent with those from the midpoint rule. Bounding the errors is done in the usual way, using the second derivatives. A look at the graph of f'' shows that $K_2 = 0.4$ is OK; thus the errors, respectively are:

$$I_1 \text{ error} \quad \le \quad \frac{0.4 \cdot 2^3}{12 \cdot 20^2} \approx 0.0007;$$

$$I_2 \text{ error} \quad \le \quad \frac{0.4 \cdot 2.5^3}{12 \cdot 20^2} \approx 0.0013.$$

Since the errors are so small, the answers are properly consistent with those from the midpoint rule.

5. When $Z = \dfrac{x - 500}{100}$, $dZ = \dfrac{dx}{100}$. Furthermore, when $x = 500$, $Z = 0$; when $x = 700$, $Z = 2$. Thus,

$$I_1 = \frac{1}{100\sqrt{2\pi}} \int_{500}^{700} \exp\left(-\frac{(x - 500)^2}{2 \cdot 100^2}\right) dx = \frac{1}{100\sqrt{2\pi}} \int_{0}^{2} \exp\left(-\frac{Z^2}{2}\right) 100\,dZ = \frac{1}{\sqrt{2\pi}} \int_{0}^{2} e^{-Z^2/2}\,dZ.$$

7. Looking at the table shows that the top 10% starts at about 1.3 standard deviations above the mean, i.e., at the raw score 630. Similarly, the top 5% starts at about $Z = 1.7$, i.e., at a raw score of 670.

9. $du = -1/x^2$ and $v = -e^{-x^2}/2$. Therefore, using integration by parts,

$$\int_{a}^{\infty} e^{-x^2}\,dx = -\frac{e^{-x^2}}{2x^2}\Bigg|_{a}^{\infty} - \frac{1}{2}\int_{a}^{\infty} \frac{e^{-x^2}}{x^2}\,dx = \frac{e^{-a^2}}{2a} - \frac{1}{2}\int_{a}^{\infty} \frac{e^{-x^2}}{x^2}\,dx.$$

10.4   l'Hôpital's rule: comparing rates

1. (a)

x	1	10	100	1000
x^2	1	100	$10,000$	1×10^6
2^x	2	1024	1.268×10^{31}	1.07×10^{301}
$(3x+10)^2$	169	1600	$96,100$	9.0601×10^6

(b) One might guess

$$\lim_{x \to \infty} \left(\frac{x^2}{2^x} \right) = 0$$

$$\lim_{x \to \infty} \left(\frac{x^2}{(3x+10)^2} \right) = \frac{1}{9}$$

$$\lim_{x \to \infty} \left(\frac{2^x}{(3x+10)^2} \right) = \infty$$

(c) Using l'Hôpital's rule:

$$\lim_{x \to \infty} \left(\frac{x^2}{2^x} \right) = \lim_{x \to \infty} \left(\frac{2x}{2^x \ln 2} \right) = \lim_{x \to \infty} \left(\frac{2}{2^x (\ln 2)^2} \right) = 0$$

$$\lim_{x \to \infty} \left(\frac{x^2}{(3x+10)^2} \right) = \lim_{x \to \infty} \left(\frac{2x}{6(3x+10)} \right) = \lim_{x \to \infty} \frac{2}{18} = \frac{1}{9}$$

$$\lim_{x \to \infty} \left(\frac{2^x}{(3x+10)^2} \right) = \lim_{x \to \infty} \left(\frac{2^x \ln 2}{6(3x+10)} \right) = \lim_{x \to \infty} \left(\frac{2^x (\ln 2)^2}{18} \right) = \infty$$

3. $\lim\limits_{x \to 0} \left(\dfrac{5x - \sin x}{x} \right) = \lim\limits_{x \to 0} \left(\dfrac{5 - \cos x}{1} \right) = 4$

5. $\lim\limits_{x \to \infty} \dfrac{e^x}{x^2 + x} = \lim\limits_{x \to \infty} \dfrac{e^x}{2x + 1} = \lim\limits_{x \to \infty} \dfrac{e^x}{2} = \infty$

7. $\lim\limits_{x \to 0} x \cot x = \lim\limits_{x \to 0} \left(\dfrac{x}{\tan x} \right) = \lim\limits_{x \to 0} \left(\dfrac{1}{\sec^2 x} \right) = 1$

9. $\lim\limits_{x \to 0^+} \left(\dfrac{\sin x}{x + \sqrt{x}} \right) = \lim\limits_{x \to 0^+} \left(\dfrac{\cos x}{1 + \frac{1}{2\sqrt{x}}} \right) = 0$

11. $\lim\limits_{x \to 0} \dfrac{\tan(3x)}{\ln(1 + x)} = \lim\limits_{x \to 0} \dfrac{3 \sec^2(3x)}{(1 + x)^{-1}} = 3$

13. $\lim\limits_{x \to \infty} \dfrac{\int_1^x \sqrt{1 + e^{-3t}}\, dt}{x} = \lim\limits_{x \to \infty} \dfrac{\sqrt{1 + e^{-3x}}}{1} = 1$ [NOTE: The improper integral in the numerator diverges to ∞ because $\sqrt{1 + e^{-3t}} > 1$ for all $t \geq 1$.]

15. (a) **Diverges** since $\cos x > 1/2$ when $0 \leq x \leq 1 \implies \displaystyle\int_0^1 \dfrac{\cos x}{x}\, dx > \dfrac{1}{2} \int_0^1 \dfrac{dx}{x} = \infty$.

(b) $\lim\limits_{x \to 0} \dfrac{\sin x}{x} = 1$ so the integrand is bounded over the entire interval of integration.

(c) $\displaystyle\int_0^\infty \sin\left(e^{-x}\right) dx = \lim_{t \to \infty} \int_0^t \sin\left(e^{-x}\right) dx = \lim_{t \to \infty} -\int_1^{e^{-t}} \sin u\, \dfrac{du}{u} = \int_0^1 \dfrac{\sin u}{u}\, du$.

$$\int_1^\infty f'(x)\,dx = \lim_{t\to\infty} f(t) - f(1).$$ Thus, the improper integral converges if (and only if) $\lim_{t\to\infty} f(t)$ is a finite number.

Since $|f(x)| \le e^{-x}\ln x$, $\lim_{t\to\infty} |f(t)| \le \lim_{t\to\infty} e^{-t}\ln t = 0$ (see problem 3d above). Thus, $\lim_{t\to\infty} f(t) = 0$.

Finally, therefore, $\int_1^\infty f'(x)\,dx = -f(1).$

$$\int_0^1 x^m(\ln x)^n\,dx = \int_\infty^0 e^{-mu}(-u)^n \left(-e^{-u}\,du\right) = (-1)^n \int_0^\infty u^n e^{(-m+1)u}\,du = (-1)^n \int_0^\infty \left(\frac{v}{m+1}\right)^n e^{-v}\,\frac{dv}{m+1} = \frac{(-1)^n}{(m+1)^{n+1}} \int_0^\infty$$

11.1 Sequences and their limits

1. $a_k = (-1/3)^k$ for $k = 0, 1, 2, \ldots$

3. $a_k = k/2^k$ for $k = 1, 2, 3 \ldots$

5. $\lim\limits_{k \to \infty} a_k$ does not exist

7. $\lim\limits_{k \to \infty} a_k = \infty$

9. $\lim\limits_{k \to \infty} a_k = \infty$

11. $\lim\limits_{k \to \infty} a_k = \infty$

13. $\lim\limits_{k \to \infty} a_k$ does not exist

15. $\lim\limits_{k \to \infty} a_k = 1$

17. $\lim\limits_{k \to \infty} a_k$ does not exist

19. $\lim\limits_{k \to \infty} a_k = \sqrt{2}$

21. $\lim\limits_{k \to \infty} a_k = 0$

23. $\lim\limits_{k \to \infty} a_k = 0$

25. $\lim\limits_{k \to \infty} a_k = 0$

27. (a) x^n *diverges* when $x > 1$ or $x \le -1$.

    (b) $\lim\limits_{n \to \infty} x^n = 0$ when $|x| < 1$.

    (c) x^n *converges* to 1 when $x = 1$.

29. $\lim\limits_{k \to \infty} a_k = 0$ when $x < 0$ and $\lim\limits_{k \to \infty} a_k = 1$ when $x = 0$.

31. $\lim\limits_{k \to \infty} u_k = 0$ when $-\infty < x < \infty$

33. The sequence is bounded above (e.g, by 0.8) and is monotone increasing.

35. Since $|\sin x| \le 1$ for all x, $\sin\left(\frac{\pi}{2^2}\right) \cdot \sin\left(\frac{\pi}{3^2}\right) \cdots \sin\left(\frac{\pi}{n^2}\right) \le \sin\left(\frac{\pi}{n^2}\right)$. Also, the inequality $0 < \sin x < x$ when $0 < x < 1$ implies that $0 < \sin\left(\pi/n^2\right) < \pi/2^n$. Therefore, since $\lim\limits_{n \to \infty} \pi/2^n = 0$, Theorem 2 implies the result.

37. Yes. The inequalities $0 < a_{n+1} = a_n/2 < a_n$ imply that the sequence is monotonically decreasing and bounded below.

39. $\lim\limits_{k \to \infty} b_k = \lim\limits_{k \to \infty} (L - a_k) = \lim\limits_{k \to \infty} L - \lim\limits_{k \to \infty} a_k = L - \lim\limits_{k \to \infty} a_k = L - L = 0$

41. $\lim\limits_{n \to \infty} \dfrac{\ln x}{n} = 0 \implies \lim\limits_{n \to \infty} e^{(\ln x)/n} = \lim\limits_{n \to \infty} x^{1/n} = 1$ for all $x > 0$.

43. $\lim\limits_{n \to \infty} a_n = \lim\limits_{n \to \infty} \dfrac{n+2}{n^3+4} = \lim\limits_{n \to \infty} \dfrac{1}{3n^2} = 0$

45. $\lim\limits_{j \to \infty} b_j = \lim\limits_{j \to \infty} \dfrac{\ln j}{\sqrt[3]{j}} = \lim\limits_{j \to \infty} \dfrac{3j^{2/3}}{j} = \lim\limits_{j \to \infty} \dfrac{3}{\sqrt[3]{j}} = 0$

47. $\lim\limits_{n\to\infty} d_n = \lim\limits_{n\to\infty} n\sin(1/n) = \lim\limits_{n\to\infty} \dfrac{\sin\left(n^{-1}\right)}{n^{-1}} = \lim\limits_{n\to\infty} \cos\left(n^{-1}\right) = 1$

49. (a) Using l'Hôpital's rule, $\lim\limits_{n\to\infty} nf(1/n) = \lim\limits_{n\to\infty} \dfrac{f(n^{-1})}{n^{-1}} = \lim\limits_{n\to\infty} \dfrac{-n^{-2}\cdot f'(n^{-1})}{-n^{-2}} = \lim\limits_{n\to\infty} f'(n^{-1}) = f'(0)$

   (b) $\lim\limits_{n\to\infty} n\arctan(1/n) = \lim\limits_{n\to\infty} \left(1 + (1/n)^2\right)^{-1} = 1$

51. (a) $a_{10} = 11/20$

   (b) $a_n = R_n$, the right Riemann sum approximation to $\displaystyle\int_0^1 x\,dx$ with n subintervals.

53. $\lim\limits_{n\to\infty} a_n = \displaystyle\int_0^1 \dfrac{dx}{1+x} = \ln 2$

11.2 Infinite series, convergence, and divergence

1. (c) When $r = 1$, $S_n = (n+1)a$ but the right-hand side of the expression is undefined (because of the zero in the denominator).

   (d) $S_n - rS_n = (a + ar + ar^2 + \cdots + ar^n) - r(a + ar + ar^2 + \cdots + ar^n) = a - ar^{n+1} = a(1 - r^{n+1})$

   (e) $S_n - rS_n = (1 - r)S_n = a(1 - r^{n+1})$

   (f) The sum is a geometric series with $a = 3$, $r = 2$, and $n = 10$. Thus, $3 + 6 + 12 + \cdots + 3072 = 6141$.

3. (a) $a_1 = 1$, $a_2 = 1/4$, $a_5 = 1/25$, $a_{10} = 1/100$

   (b) $S_1 = 1$; $S_2 = 5/4$; $S_5 = 5269/3600 \approx 1.46361$; $S_{10} = 1968329/1270080 \approx 1.54977$

   (c) S_n is an increasing sequence because the terms of the series are all positive (i.e., $a_k > 0$ for all k).

   (d) $R_1 = \pi^2/6 - S_1 \approx \ldots$ $R_2 = \pi^2/6 - S_2 \approx \ldots$ $R_5 = \pi^2/6 - S_5 \approx \ldots$ $R_{10} = \pi^2/6 - S_{10} \approx \ldots$

   (e) Yes because $R_n = \pi^2/6 - S_n$ and $S_{n+1} > S_n$.

   (f) $R_{20} \approx 0.048771$. Since $0 < R_{n+1} < R_n$ for all $n \geq 1$, $0 < R_n < 0.05$ for all $n \geq 20$.

   (g) $\lim\limits_{n \to \infty} R_n = 0$

5.

7. (a) $S_1 = 8/15$, $S_2 = 103/165$, $S_5 = 2626616/3892119$, $S_{10} = 2923789575132170780644253/43193822921846261229\cdots$

   (b) The sequence of partial sums is increasing because each term of the series is a positive number (i.e., $S_{n+1} = S_n + a_{n+1} > S_n$ because $a_{n+1} > 0$).

      Since $a_j < 3^{-j}$ for all $j \geq 0$, $S_n = \sum\limits_{j=0}^{n} a_j < \sum\limits_{j=0}^{n} 3^{-j} = \frac{1}{2}(3 - 3^{-n}) < 3$ for all $n \geq 0$. Thus, each term in the sequence of partial sums is bounded above by 3..

   (c) Since the sequence of partial sums is increasing and bounded above, it must converge. This implies that the infinite series converges.

9. (a) $\dfrac{1}{4} + \dfrac{1}{16} + \dfrac{1}{36} + \dfrac{1}{64} + \dfrac{1}{100} + \cdots = \dfrac{1}{4}\left(1 + \dfrac{1}{4} + \dfrac{1}{9} + \cdots\right) = \dfrac{\pi^2}{24}$.

   (b) $\sum\limits_{k=0}^{\infty} \dfrac{1}{(2k+1)^2} = \sum\limits_{k=1}^{\infty} \dfrac{1}{k^2} - \dfrac{1}{4}\sum\limits_{k=1}^{\infty} \dfrac{1}{k^2} = \dfrac{\pi^2}{6} - \dfrac{\pi^2}{24} = \dfrac{\pi^2}{8}$.

   (c) $\sum\limits_{m=1}^{\infty} \dfrac{(-1)^{m+1}}{m^2} = \sum\limits_{m=1}^{\infty} \dfrac{1}{m^2} - \dfrac{1}{2}\sum\limits_{m=1}^{\infty} \dfrac{1}{m^2} = \dfrac{\pi^2}{12}$.

      Alternatively, $\sum\limits_{m=1}^{\infty} \dfrac{(-1)^{m+1}}{m^2} = \sum\limits_{k=0}^{\infty} \dfrac{1}{(2k+1)^2} - \sum\limits_{j=1}^{\infty} \dfrac{1}{(2j)^2} = \dfrac{\pi^2}{8} - \dfrac{\pi^2}{24} = \dfrac{\pi^2}{12}$.

11. $2 - 5 + 9 + \dfrac{1}{3} + \dfrac{1}{9} + \dfrac{1}{27} + \dfrac{1}{81} + \cdots + \dfrac{1}{3^n} + \cdots = 6 + \dfrac{1}{3}\sum\limits_{j=0}^{\infty}\left(\dfrac{1}{3}\right)^j = 6 + \sum\limits_{j=1}^{\infty}\left(\dfrac{1}{3}\right)^j = 6 + 1/2 = 6.5$

13. $\sum\limits_{k=3}^{\infty}\left(\dfrac{e}{\pi}\right)^k = \left(\dfrac{e}{\pi}\right)^3 \sum\limits_{k=0}^{\infty}\left(\dfrac{e}{\pi}\right)^k = \left(\dfrac{e}{\pi}\right)^3 \dfrac{1}{1 - e/\pi} = \dfrac{e^3}{\pi^2(\pi - e)} \approx 4.8076$.

15. $\sum\limits_{i=10}^{\infty}\left(\dfrac{2}{3}\right)^i = \left(\dfrac{2}{3}\right)^{10} \sum\limits_{i=0}^{\infty}\left(\dfrac{2}{3}\right)^i = \left(\dfrac{2}{3}\right)^{10} \dfrac{1}{1 - (2/3)} = \dfrac{1024}{19683} \approx 0.052025$

17. $\sum\limits_{j=0}^{\infty} \dfrac{3^j + 4^j}{5^j} = \sum\limits_{j=0}^{\infty} \dfrac{3^j}{5^j} + \sum\limits_{j=0}^{\infty} \dfrac{4^j}{5^j} = 5/2 + 5 = 15/2$

19. The series converges when $-1 < x/5 < 1$ (i.e., when $-5 < x < 5$). $\displaystyle\sum_{m=2}^{\infty}\left(\frac{x}{5}\right)^m = \left(\frac{x}{5}\right)^2 \sum_{m=0}^{\infty}\left(\frac{x}{5}\right)^m = \left(\frac{x}{5}\right)^2 \frac{1}{1-(x/5)} = \frac{x^2}{5(5-x)}$.

21. Since $\displaystyle\sum_{k=1}^{\infty} x^{-k} = \sum_{k=1}^{\infty}\left(x^{-1}\right)^k$, the series converges when $\left|x^{-1}\right| < 1$. Therefore, the series converges when $x < -1$ or $x > 1$.

$\displaystyle\sum_{k=1}^{\infty} x^{-k} = \frac{1}{x}\sum_{k=0}^{\infty} x^{-k} = \frac{1}{x}\frac{1}{1-x^{-1}} = \frac{1}{x-1}$.

23. The series converges when $\left|\dfrac{1}{1-x}\right| < 1$. Thus, the series converges when $x < 0$ or $x > 2$. $\displaystyle\sum_{j=4}^{\infty}\frac{1}{(1-x)^j} = -\frac{1}{x(1-x)^3} = \frac{1}{x(x-1)^3}$.

25. $a_{n+1} = 4 - \displaystyle\sum_{k=1}^{n}(1/2)^k \implies \lim_{n\to\infty} a_n = 3$

27. $S_n = 1 - 1/n \implies \lim_{n\to\infty} S_n = 1$. Thus, the series converges to 1.

29. $S_n = \ln(n+1)$. Since $\lim_{n\to\infty} S_n = \infty$, the series diverges.

31. $S_n = \dfrac{3}{4} + \dfrac{1}{4}(-1)^{n+1} - \dfrac{1}{2}(-1)^{n+1}(n+1)$. Since $\lim_{n\to\infty} S_n$ doesn't exist, the series diverges.

33. The series $\displaystyle\sum_{k=0}^{\infty}\frac{3}{10}\left(-\frac{1}{2}\right)^k$ converges to $1/5$.

35. Diverges by the nth term test.

37. Diverges by the nth term test: $\displaystyle\lim_{n\to\infty}\frac{n+1}{2n+1} = \frac{1}{2} \neq 0$.

39. Diverges by the nth term test. (Since $0 < \pi - e$, $\displaystyle\lim_{k\to\infty} k^{\pi-e} = \infty$.)

41. $\displaystyle\sum_{k=1}^{\infty}\left(\int_k^{k+1}\frac{dx}{x^2}\right) = \sum_{k=1}^{\infty}\frac{1}{k(k+1)} = \lim_{n\to\infty}\left(1 - \frac{1}{n+1}\right) = 1$ Alternatively, $\displaystyle\sum_{k=1}^{\infty}\left(\int_k^{k+1}\frac{dx}{x^2}\right) = \int_1^{\infty}\frac{dx}{x^2} = 1$.

43. Diverges by the nth term test: $\displaystyle\lim_{n\to\infty}\sqrt[n]{\pi} = 1 \neq 0$,

45. (a) $\displaystyle\lim_{k\to\infty}\frac{1}{\sqrt{k}} = 0$

    (b) Since $1/\sqrt{n} \leq 1/\sqrt{k}$ when $1 \leq k \leq n$, $S_n = \displaystyle\sum_{k=1}^{n}\frac{1}{\sqrt{k}} \geq \sum_{k=1}^{n}\frac{1}{\sqrt{n}} = \frac{n}{\sqrt{n}} = \sqrt{n}$.

    (c) It follows from part (a) that $\displaystyle\lim_{n\to\infty} S_n = \infty$.

47. (a) $\displaystyle\lim_{n\to\infty} S_n = \ln 2$.

    (b) Since the sequence of its partial sums has a limit, the series converges.

49. (a) $a_k = (-1)^{k+1}$

    (b) By definition, the infinite series $\displaystyle\sum_{k=1}^{\infty} a_k$ converges if the sequence of its partial sums has a limit—that is, the series converges if $\displaystyle\lim_{n\to\infty} S_n$ exists. Since $a_k \geq 0$, S_n is an increasing sequence bounded above by 100 so Theorem 3 implies that this sequence has a limit and, therefore, that the infinite series converges.

(c) $\displaystyle\sum_{k=1}^{\infty} a_k = \sum_{k=1}^{10^6-1} a_k + \sum_{k=10^6}^{\infty} a_k.$ $\displaystyle\sum_{k=1}^{10^6-1} a_k$ is a finite sum, so it is a real number. The infinite series $\displaystyle\sum_{k=10^6}^{\infty} a_k$ converges because the sequence of its partial sums is increasing and bounded above. Thus, the original infinite series converges.

51.

53. (a) If the interval $[1, n+1]$ is divided into n equal subintervals, each subinterval has length one and the left endpoint of the jth subinterval is $1/j$. Thus, $L_n = \displaystyle\sum_{j=1}^{n} 1/j.$

(b) Since $1/x$ is a decreasing function on the interval $[1, \infty)$, $L_n > I_n$ for every $n \geq 1$. Now, $I_n = \ln(n+1) \implies \displaystyle\lim_{n\to\infty} I_n = \infty$ = Thus, since $H_n = L_n$, the sequence of partial sums of the harmonic series does not have a finite limit (i.e., the harmonic series diverges).

55. (a) When the interval $[1, 2]$ is divided into n equal subintervals, each subinterval has length $1/n$ and the left endpoint of the kth subinterval is $1 + (k-1)/n = (n+k-1)/n$. Thus, $L_n = \dfrac{1}{n}\displaystyle\sum_{k=1}^{n} \dfrac{n}{n+k-1} = \sum_{k=1}^{n} \dfrac{1}{n+k-1} = \sum_{k=0}^{n-1} \dfrac{1}{n+k}.$

(b) Since $1/x$ is a decreasing function on the interval $[1, 2]$, $L_n > I = \ln 2$ for all $n \geq 1$.

(c) $H_{2n} - H_n = \displaystyle\sum_{k=1}^{2n} \dfrac{1}{k} - \sum_{k=1}^{n} \dfrac{1}{k} = \sum_{k=n+1}^{2n} \dfrac{1}{k} = \dfrac{1}{n+1} + \dfrac{1}{n+2} + \cdots + \dfrac{1}{2n-1} + \dfrac{1}{2n} = \left(L_n - \dfrac{1}{n}\right) + \dfrac{1}{2n} = L_n - \dfrac{1}{2n}$

(d) $\displaystyle\lim_{n\to\infty} (H_{2n} - H_n) = \lim_{n\to\infty}\left(L_n - \dfrac{1}{2n}\right) = \ln 2 - 0 = \ln 2$

(e) The result in part (d) implies that the partial sums of the harmonic series do not have a finite limit.

(f) Since $1/(2n) \leq 1/2$ for all $n \geq 1$, $H_{2n} - H_n = L_n - 1/(2n) \geq L_n - 1/2$ for all $n \geq 1$. Since, by part (b), $L_n > \ln 2$ for all $n \geq 1$, $H_{2n} - H_n > \ln 2 - 1/2$ for all $n \geq 1$.

(g) $H_2 = H_1 + L_1 - \tfrac{1}{2} = 1 + L_1 - \tfrac{1}{2} > 1 + \ln 2 - \tfrac{1}{2}$

(h) By part (g), the formula holds when $m = 1$. To show that the formula holds for all $m \geq 1$, we use induction: Assume that the formula holds for $m - 1$. That is, assume that $H_{2^{m-1}} > 1 + (m-1)\left(\ln 2 - \tfrac{1}{2}\right)$. Then, by part (f),

$H_{2^m} - H_{2^{m-1}} > \ln 2 - \tfrac{1}{2}$ so $H_{2^m} > H_{2^{m-1}} + \left(\ln 2 - \tfrac{1}{2}\right) > 1 + (m-1)\left(\ln 2 - \tfrac{1}{2}\right) + \left(\ln 2 - \tfrac{1}{2}\right) = 1 + m\left(\ln 2 - \tfrac{1}{2}\right).$ Therefore, the formula holds for all $m \geq 1$.

Testing for convergence; estimating limits

(a) When $k \geq 0$, $k + 2^k \geq 2^k \implies a_k \leq 1/2^k = 2^{-k}$. Since $\displaystyle\sum_{k=0}^{\infty} 2^{-k}$ converges, the comparison test implies that $\displaystyle\sum_{k=0}^{\infty} a_k$ converges.

(b) $R_{10} = \displaystyle\sum_{k=11}^{\infty} a_k < \sum_{k=11}^{\infty} 2^{-k} = 2^{-11} \sum_{k=0}^{\infty} 2^{-k} = 2^{-10}$

(c) Since $R_{10} < 2^{-10} \approx 0.00097656$, S_n has the desired accuracy if $n \geq 10$. $S_{10} = \displaystyle\sum_{k=0}^{10} a_k = \frac{127807216183}{75344540040} \approx 1.6963$.

(d) Since the terms of the series are all positive, the estimate in part (c) *underestimates* the limit.

(a) Let $M = \displaystyle\sum_{k=1}^{\infty} b_k = \lim_{n\to\infty} T_n$. Since $0 \leq a_k \leq b_k$ for all $k \geq 1$, $S_n \leq T_n$ for all $n \geq 1$. Furthermore, since $0 \leq b_k$ for all
$k \geq 1$, $\{T_n\}$ is an increasing sequence whose limit is M. It follows that $S_n \leq T_n \leq M$ for every $n \geq 1$.

(b) S_n is an increasing sequence because $0 \leq a_k$ for all $k \geq 1$: $S_{n+1} = S_n + a_{n+1} > S_n$.

(c) Together, parts (a) and (b) imply that $\{S_n\}$ is an increasing sequence that is bounded above, so it converges. Since $\{S_n\}$
is a sequence of partial sums of the infinite series $\displaystyle\sum_{k=1}^{\infty} a_k$, convergence of the sequence implies convergence of the series.

(d) Divergence of the series implies that the sequence of its partial sums diverges. Since $0 \leq a_k$ for all $k \geq 1$, $S_n \geq 0$ for
all $n \geq 1$ and $\{S_n\}$ is an increasing sequence. Such a sequence diverges only if $\displaystyle\lim_{n\to\infty} S_n = \infty$.

(e) Since $S_n \leq T_n$ for all $n \geq 1$, $\displaystyle\lim_{n\to\infty} S_n = \infty \implies \lim_{n\to\infty} T_n = \infty$.

$$\int_{n+1}^{\infty} a(x)\,dx < \sum_{k=n+1}^{\infty} a_k < \int_{n}^{\infty} a(x)\,dx$$

(a) $S_{n+1} = \displaystyle\sum_{k=1}^{n+1} a_k = S_n + a_k \geq S_n$ since $a_k = a(k) \geq 0$ for all integers $k \geq 1$.

(b) $\displaystyle\int_{1}^{\infty} a(x)\,dx = \int_{1}^{n} a(x)\,dx + \int_{n}^{\infty} a(x)\,dx \implies \int_{1}^{n} a(x)\,dx \leq \int_{1}^{\infty} a(x)\,dx$ because $a(x) \geq 0$ for all $x \geq 1 \implies \int_{n}^{\infty} a(x)\,dx$

(c) Part (a) implies that $\{S_n\}$ is an increasing sequence. Since $S_n \leq \displaystyle\int_{1}^{n} a(x)\,dx$, part (b) implies that the sequence of partial

sums is bounded above by $\displaystyle\int_{1}^{\infty} a(x)\,dx$. Thus, the sequence of partial sums converges to a limit.

$$\int_{1}^{\infty} \frac{dx}{x^{3/2}} = -\frac{2}{\sqrt{x}}\Big|_{1}^{\infty} = 2 \implies \sum_{k=1}^{\infty} \frac{1}{k\sqrt{k}} \text{ converges and } 2 \leq \sum_{k=1}^{\infty} \frac{1}{k\sqrt{k}} \leq 3.$$

(a) Let $a_j = \dfrac{j^2}{j!}$. Since $\displaystyle\lim_{j\to\infty} \frac{a_{j+1}}{a_j} = \lim_{j\to\infty} \frac{\frac{(j+1)^2}{(j+1)!}}{\frac{j^2}{j!}} = \lim_{j\to\infty} \frac{(j+1)^2}{j^2} \cdot \frac{j!}{(j+1)!} = \lim_{j\to\infty} \frac{j+1}{j^2} = 0 < 1$, $\displaystyle\sum_{j=0}^{\infty} a_j$ converges.

(b) $\displaystyle\lim_{k\to\infty} \frac{\frac{2^{k+1}}{(k+1)!}}{\frac{2^k}{k!}} = \lim_{k\to\infty} \frac{2}{k+1} = 0 \implies \sum_{k=1}^{\infty} \frac{2^k}{k!}$ converges.

(a) The following inequalities are apparent from the figures on pp. 241–243: $\displaystyle\int_{1}^{n+1} a(x)\,dx < \sum_{k=1}^{n} a_k < a_1 + \int_{1}^{n} a(x)\,dx$.

Taking $a(x) = 1/x$, these inequalities imply that $\ln(n+1) < H_n < 1 + \ln n$.

(b) Since $H_N < 1 + \ln N$, $H_N > 10 \implies 1 + \ln N > 10$. From this it follows that $N \geq 8104$.

(c) Using the inequalities derived in part (a), $\dfrac{\ln(n+1)}{\ln n} < \dfrac{H_n}{\ln n} < \dfrac{1 + \ln n}{\ln n}$. Since $\lim\limits_{n \to \infty} \dfrac{\ln(n+1)}{\ln n} = 1$ and $\lim\limits_{n \to \infty} \dfrac{1 + \ln n}{\ln n} = 1$,

$\lim\limits_{n \to \infty} \dfrac{H_n}{\ln n} = 1$ (Theorem 2).

(d) First, observe that $a_n - a_{n+1} = (H_n - \ln n) - (H_{n+1} - \ln(n+1)) = \ln(n+1) - \ln n - \dfrac{1}{n+1}$. Then, note that

$\displaystyle \int_n^{n+1} x^{-1}\, dx = \ln(n+1) - \ln n > \dfrac{1}{n+1}$ since x^{-1} is a decreasing function on the interval $[n, n+1]$. Thus, $a_n - a_{n+1} > 0$.

(e) The sequence a_n is decreasing and bounded below by 0 (since $H_n - \ln n > \ln(n+1) - \ln n > 0$). Thus, it converges (Theorem 3).

17. (a) The integral test can't be used to prove that the series converges because the function $a(x) = \dfrac{2 + \sin x}{x^2}$ is not decreasing on the interval $[1, \infty)$.

(b) Since $0 < \dfrac{2 + \sin k}{k^2} \leq \dfrac{3}{k^2}$ for all $k \geq 1$, and $\displaystyle\sum_{k=1}^{\infty} \dfrac{3}{k^2}$ converges, the comparison test implies that the series $\displaystyle\sum_{k=1}^{\infty} \dfrac{2 + \sin k}{k^2}$ converges.

19. (a) For $m = 1, 2, 3, \ldots$, $\dfrac{a_{2m}}{a_{2m-1}} = \left(\dfrac{2}{3}\right)^m$ and $\dfrac{a_{2m+1}}{a_{2m}} = \dfrac{1}{2}\left(\dfrac{3}{2}\right)^m$.

(b) Because the limit in part (a) does not exist, the ratio test says nothing about the convergence of the series $\displaystyle\sum_{k=1}^{\infty} a_k$.

(c) $\displaystyle\sum_{k=1}^{\infty} a_k = \sum_{k=1}^{\infty} 2^{-k} + \sum_{k=1}^{\infty} 3^{-k} = 1 + \dfrac{1}{2} = \dfrac{3}{2}$

21. $\displaystyle\sum_{n=1}^{\infty} \dfrac{1}{n^2 + \sqrt{n}} < \sum_{k=1}^{\infty} \dfrac{1}{n^2} \leq 1 + \int_1^{\infty} \dfrac{dx}{x^2} = 2.$

23. $\displaystyle\sum_{m=1}^{\infty} \dfrac{1}{m\sqrt{1 + m^2}} < \sum_{m=1}^{\infty} \dfrac{1}{m^2} \leq 2.$

25. Diverges. Each term of this series is a constant multiple $(1/100)$ of the corresponding term of the harmonic series.

27. Converges—comparison test: $\displaystyle\sum_{m=1}^{\infty} \dfrac{m^3}{m^5 + 3} < \sum_{m=1}^{\infty} \dfrac{m^3}{m^5} = \sum_{m=1}^{\infty} \dfrac{1}{m^2} \leq 2.$

29. Diverges. Each term of the series is a constant multiple $\left(\frac{1}{\ln 10}\right)$ of the corresponding term of the harmonic series since $\ln\left(10^k\right) = k \ln 10$.

31. Diverges—nth term test: $\lim\limits_{k \to \infty} \dfrac{k^2}{5k^2 + 3} = \dfrac{1}{5} \neq 0.$

33. Diverges—comparison test: $\displaystyle\sum_{n=2}^{\infty} \dfrac{1}{\sqrt[3]{n^2 - 1}} > \sum_{n=2}^{\infty} \dfrac{1}{n^{2/3}}.$

35. Converges—integral test: $\displaystyle\int_0^{\infty} e^{-x^2}\, dx = \sqrt{\pi}/2 \implies \sum_{m=0}^{\infty} e^{-m^2} \leq 1 + \int_0^{\infty} e^{-x^2}\, dx = 1 + \sqrt{\pi}/2.$

37. Diverges—comparison test: $\displaystyle\sum_{k=1}^{\infty} \dfrac{k!}{(k+1)! - 1} > \sum_{k=1}^{\infty} \dfrac{k!}{(k+1)!} = \sum_{k=1}^{\infty} \dfrac{1}{k+1} = \dfrac{1}{2} + \dfrac{1}{3} + \dfrac{1}{4} + \cdots$

39. Diverges by the nth term test: $\displaystyle\lim_{n\to\infty}\sum_{k=1}^{n}k^{-1}=\infty$

41. The comparison $\displaystyle\sum_{k=1}^{\infty}\frac{k}{k^6+17}<\sum_{k=1}^{\infty}\frac{k}{k^6}=\sum_{k=1}^{\infty}\frac{1}{k^5}$ shows that the original series converges.

Since $R_n\le\displaystyle\int_n^\infty\frac{x}{x^6+17}\,dx<\int_n^\infty\frac{dx}{x^5}=\frac{1}{4n^4}<0.005$ when $n\ge 3$, the estimate $L\approx\displaystyle\sum_{k=1}^{3}\frac{k}{k^6+17}=\frac{2546}{30213}\approx 0.084268$
is guaranteed to be in error by no more than 0.005.

43. (a) The assumption that $a(x)$ is decreasing is necessary to ensure that the desired geometric relationships hold (e.g., that
$\displaystyle\int_1^{n+1}a(x)\,dx\le\sum_{k=1}^{n}a_k$).

(b) Under the new assumption the inequality in the second bullet must be replaced by $\displaystyle\int_{10}^\infty a(x)\,dx\le\sum_{k=10}^{\infty}a_k\le a_{10}+\int_{10}^\infty a(x)\,dx$
and the condition $n\ge 10$ must be placed on the inequality in the last bullet. No other changes in the conclusion of the theorem are necessary.

45. (a) $k!=\overbrace{k\cdot(k-1)\cdot\ldots\cdot 2}^{(k-1)\text{-terms}}\ge\overbrace{2\cdot 2\cdot\ldots\cdot 2}^{(k-1)\text{-terms}}=2^{k-1}\implies\dfrac{1}{k!}\le\dfrac{1}{2^{k-1}}$

(b) $k!=\overbrace{k\cdot(k-1)\cdot\ldots\cdot(k-10)\cdot 10\cdot\ldots\cdot 2}^{(k-10)\text{-terms}}\,10!\ge\overbrace{10\cdot 10\cdot\ldots\cdot 10}^{(k-10)\text{-terms}}\cdot 10!=10^{k-10}\cdot 10!\implies\dfrac{1}{k!}\le\dfrac{1}{10!\,10^{k-10}}$

(c) Let $S_n=\displaystyle\sum_{k=0}^{n}\frac{1}{k!}$. S_{10} underestimates $\displaystyle\sum_{k=0}^{\infty}\frac{1}{k!}$ since S_n is a monotonically increasing sequence.

$R_{10}=\displaystyle\sum_{k=0}^{\infty}\frac{1}{k!}-S_{10}=\sum_{k=11}^{\infty}\frac{1}{k!}\le\sum_{k=11}^{\infty}\frac{1}{10!\,10^{k-10}}=\frac{1}{10!}\sum_{k=1}^{\infty}\frac{1}{10^k}=\frac{1}{10!}\cdot\frac{1}{9}=\frac{1}{32659200}\approx 3.0619\times 10^{-8}$

47. There does not exist a number $r<1$ such that $\dfrac{a_{k+1}}{a_k}\le r$ for all $k\ge 1$.

11.4   Absolute convergence; alternating series

1. (a) The series converges conditionally. (After the fifth term, the series has the same terms as the alternating harmonic series.)

   (b) $S_{15} = 1 + 2 + 3 + 4 + 5 + \displaystyle\sum_{k=6}^{15} \frac{(-1)^{k+1}}{k} = 15 - \frac{20887}{360360} \approx 14.942$. S_{15} *overestimates* S because the last term included
   in the alternating series was positive.

   (c) $14.902 < S < 14.902 + \frac{1}{61} \approx 14.918$

   (d) $S = 15 + \left(\ln 2 - \displaystyle\sum_{k=1}^{5} \frac{(-1)^{k+1}}{k} \right) = 15 + \ln 2 - \frac{47}{60} \approx 14.910$

3. Let $b_k = a_{k+10^9}$. The alternating series test can be used to show that $\displaystyle\sum_{k=1}^{\infty} (-1)^{k+1} b_k$ converges. Since

$$\sum_{k=1}^{\infty} (-1)^{k+1} a_k = \sum_{k=1}^{10^9} (-1)^{k+1} a_k + \sum_{k=10^9+1}^{\infty} (-1)^{k+1} a_k = \sum_{k=1}^{10^9} (-1)^{k+1} a_k + \sum_{k=1}^{\infty} (-1)^{k+1} b_k, \text{ the series } \sum_{k=1}^{\infty} (-1)^{k+1} a_k \text{ converges.}$$

5. converges absolutely—$\displaystyle\sum_{j=1}^{\infty} \frac{1}{j^2}$ is a convergent p-series $(p = 2)$

$$\frac{3}{4} < \sum_{j=1}^{\infty} \frac{(-1)^{j+1}}{j^2} < 1$$

7. converges conditionally—$\displaystyle\sum_{k=4}^{\infty} \frac{\ln k}{k}$ diverges by the integral test but the terms of the series form a decreasing sequence and

$$\lim_{k \to \infty} \frac{\ln k}{k} = 0$$

$$\frac{\ln 4}{4} - \frac{\ln 5}{5} < \sum_{k=4}^{\infty} (-1)^k \frac{\ln k}{k} < \frac{\ln 4}{4}$$

9. converges conditionally—$\displaystyle\sum_{n=1}^{\infty} \frac{\cos(n\pi)}{n} = \sum_{n=1}^{\infty} \frac{(-1)^n}{n}$ which is (almost) the alternating harmonic series

$$-1 < \sum_{n=1}^{\infty} \frac{\cos(n\pi)}{n} = -\ln 2 < -1/2$$

11. converges absolutely—Let $a_m = 4m^3/2^m$. Then $\displaystyle\lim_{m \to \infty} \frac{a_{m+1}}{a_m} = \lim_{m \to \infty} \frac{(m+1)^3}{2m^3} = \frac{1}{2} < 1$ so $\displaystyle\sum_{m=0}^{\infty} \frac{4m^3}{2^m}$ converges by the ratio
   test.

   The terms of the series are decreasing in absolute value for all $m \geq 4$. Thus,

$$\sum_{m=0}^{5} (-1)^m a_m = -\frac{57}{8} < \sum_{m=0}^{\infty} (-1)^m \frac{4m^3}{2^m} < \sum_{m=0}^{4} (-1)^m a_m = \frac{17}{2}$$

13. converges absolutely—$\displaystyle\lim_{j \to \infty} \frac{a_{j+1}}{a_j} = \lim_{j \to \infty} \frac{j+1}{(j^2 + 2j + 1) \cdots (j^2 + 1)} = 0$

   The terms of the series are decreasing in absolute value for all $j \geq 1$. Thus, $\displaystyle\sum_{j=0}^{1} (-1)^j a_j < \sum_{j=0}^{\infty} (-1)^j a_j < \sum_{j=0}^{2} (-1)^j a_j = \frac{1}{12}$.

15. The series converges absolutely by the integral test. $\left| L - \sum_{k=1}^{N} a_k \right| \leq 0.005$ when $N \geq 4$ since $3/5^4 < 0.005$. Using $N = 4$,

$L \approx -\dfrac{19615}{6912} \approx -2.8378.$

17. The series converges absolutely by the comparison test using $b_k = (2/7)^k$. $\left| L - \sum_{k=0}^{N} a_k \right| \leq 0.005$ when $N \geq 4$ since

$2^5/(7^5 + 5) < 0.005$. Using $N = 4$, $L \approx \dfrac{68917177}{84877260} \approx 0.81196.$

19. The series converges absolutely by the ratio test. $\left| L - \sum_{k=5}^{N} a_k \right| \leq 0.005$ when $N \geq 13$ since $14^{10}/10^{14} < 0.005$. Using

$N = 13$, $L \approx -\dfrac{573982077919709}{10000000000000} \approx -57.398.$

21. $a_k = (-1)^k/\sqrt{k}$

23. No. This would contradict Theorem 9.

25. (a) Let $a_k = (-1)^{k+1}/k^2$ and $b_k = 1/k^2$. Then $a_{2m-1} = b_{2m-1}$ and $a_{2m} < b_{2m}$ for $m = 1, 2, 3, \ldots$. Thus,

$\sum_{k=n+1}^{\infty} a_k \leq \sum_{k=n+1}^{\infty} b_k.$

(b) According to part (a), for each n the tail of the alternating series is smaller than the tail of the series of positive terms. Thus, the error made by approximating the alternating series by its nth partial sum is less than the error made by approximating the series with positive terms by its nth partial sum.

11.5 Power series

1.

3. $\lim\limits_{n\to\infty} \left| \dfrac{\frac{x^{n+1}}{\sqrt{n+1}}}{\frac{x^n}{\sqrt{n}}} \right| = \lim\limits_{n\to\infty} \dfrac{\sqrt{n}}{\sqrt{n+1}} \cdot |x| < 1$ when $|x| < 1$. Thus, the radius of convergence is $R = 1$.

5. $\left| \dfrac{x^n}{n!+n} \right| \le \dfrac{|x|^n}{n!} \implies R = \infty$

7. $\left| \dfrac{a_{n+1}}{a_n} \right| = \left(\dfrac{n+1}{n} \right)^n \cdot (n+1) \cdot |x| \implies R = 0$

9. $R = 1$

11. $R = 1$

13. For each series in this problem, the ratio test can be used to prove that the series converges when $-R < x < R$.

   (a) The nth term test can be used to prove that the series diverges when $|x| > R$.

   (b) The alternating series test can be used to show that the series converges when $x = -R$. When $x = R$ the series becomes the harmonic series (i.e., it diverges).

   (c) When $x = R$ the series becomes $\sum\limits_{k=1}^{\infty} \dfrac{1}{k^2}$ which converges absolutely.

   (d) $\sum\limits_{k=1}^{\infty} \dfrac{(-x)^k}{kR^k} = \sum\limits_{k=1}^{\infty} (-1)^k \dfrac{x^k}{kR^k}$

15. (a) By definition, the radius of convergence of a power series is the largest value of R such that the series converges for all x such that $|x| < R$.

   (b) This power series converges when $|x - 1| < 2$ and diverges when $|x - 1| > 2$. Thus, its radius of convergence $R = 2$.

   (c) Let $z = x - 3$. Since $\sum\limits_{k=0}^{\infty} z^k$ converges only when $-2 < z \le 2$, $\sum\limits_{k=0}^{\infty} a_k (x - 3)^k$ converges only when $1 < x \le 5$.

   (d) The power series $\sum\limits_{k=0}^{\infty} a_k (x + 1)^k$ converges only when $-2 < x + 1 \le 2$. Thus, its interval of convergence is $(-3, 1]$.

17. $(1, 5)$

19. $[-3, 5)$

21. $[-6, -4]$

23. The information given implies that the power series converges on the interval $[-3, 3)$, diverges when $x \ge 7$, and diverges when $x < -7$. It does not imply anything about convergence or divergence on the intervals $[-7, -3)$ and $[3, 7)$.

   (a) cannot

   (b) might

   (c) might

   (d) cannot

   (e) might

   (f) might

25. (a) Cannot be true. The interval of convergence of a power series is symmetric around and includes its base point ($b = 1$ in this case).

(b) Might be true. (The statement is true when $a_k = 1/k!$ but it is false when $a_k = 1$.)

(c) Must be true. If the radius of convergence of the power series is 3, then the interval of convergence includes all values of x such that $|x - 1| < 3$.

(d) Cannot be true. The interval of convergence of this power series must be symmetric about the point $b = 1$.

(e) Cannot be true. The interval of convergence of the power series is the solution set of the inequality $|x - 1| < 8$. Thus, the radius of convergence of the power series is 8.

27. (a) The series $\displaystyle\sum_{n=0}^{\infty} \frac{2 \cdot 10^n}{3^n + 5}$ diverges by the ratio test—the limit of the ratio of successive terms of the series is $10/3 > 1$.

(b) The power series defining f converges when $-3 < x < 3$. Thus, only 0.5 and 1.5 are in the domain of f.

(c) $\displaystyle f(1) - \sum_{n=0}^{N} \frac{2}{3^n + 5} = \sum_{n=N+1}^{\infty} \frac{2}{3^n + 5} < \sum_{n=N+1}^{\infty} \frac{2}{3^n} = \frac{1}{3^{N+1}} \sum_{n=0}^{\infty} \frac{2}{3^n} = \frac{1}{3^{N+1}} \cdot 3 = \frac{1}{3^N}$. Thus, since $3^{-5} < 0.01$,

$\displaystyle\sum_{n=0}^{5} \frac{2}{3^n + 5} = \frac{367273}{447888} \approx 0.82001$ approximates $f(1)$ within 0.01.

29. (a) The domain of g is $[-9, 1]$.

(b) $\displaystyle g(0) - \sum_{n=1}^{N} \frac{4^n}{n^3 5^n} = \sum_{n=N+1}^{\infty} \frac{4^n}{n^3 5^n} < \int_{N}^{\infty} \frac{dx}{x^3} = \frac{1}{2N^2} \le 0.005$ when $N \ge 10$. Thus,

$$g(0) \approx \sum_{n=0}^{10} \frac{4^n}{n^3 5^n} = \frac{277892997449134}{305233154296875} \approx 0.91043.$$

(c) The approximation $g(-5) \approx -\dfrac{1}{5}$ is correct within 0.005 because the series defining $g(-5)$ is an alternating series. Since the magnitude of the second term in the series is 0.005, the error made by approximating the series by its first term is smaller than 0.005.

11.6   Power series as functions

1.

3. $f(x) = \left(1 - x^2\right)^{-1} = \sum_{k=0}^{\infty} x^{2k}$ [Substitute $u = x^2$ into the power series representation of $(1 - u)^{-1}$.]

5. $f(x) = \dfrac{x}{1 - x^4} = x \sum_{k=0}^{\infty} x^{4k} = \sum_{k=0}^{\infty} x^{4k+1}$

7. $\cos(x^2) = \sum_{k=0}^{\infty} (-1)^k \dfrac{x^{4k}}{(2k)!}$

9. $\dfrac{1}{2 + x} = \dfrac{1}{2} \left(\dfrac{1}{1 + (x/2)} \right) = \dfrac{1}{2} \sum_{k=0}^{\infty} (-1)^k \left(\dfrac{x}{2} \right)^k$

$\dfrac{x^{2k}}{(2k)!}$

11. $\sin x + \cos x = \sum_{k=0}^{\infty} (-1)^k \left(\dfrac{x^{2k}}{(2k)!} + \dfrac{x^{2k+1}}{(2k+1)!} \right)$

13. $\ln\left(1 + x^2\right) = \sum_{k=1}^{\infty} (-1)^{k+1} \dfrac{x^{2k}}{k}$

15. $\ln\left(\dfrac{1 + x}{1 - x} \right) = \ln(1 + x) - \ln(1 - x) = 2 \sum_{k=0}^{\infty} \dfrac{x^{2k+1}}{2k + 1}$

17. *This is the same problem as #14 above.*

19. $\dfrac{\cos x}{1 + x^2} = \left(\sum_{k=0}^{\infty} (-1)^k \dfrac{x^{2k}}{(2k)!} \right) \left(\sum_{k=0}^{\infty} (-1)^k x^{2k} \right) = 1 - \dfrac{3}{2}x^2 + \dfrac{37}{24}x^4 - \dfrac{1111}{720}x^6 + \cdots$

21. $\arctan x \sin(4x) = \left(\sum_{k=0}^{\infty} (-1)^k \dfrac{x^{2k+1}}{2k + 1} \right) \left(\sum_{k=0}^{\infty} (-1)^k \dfrac{x^{2k+1}}{(2k + 1)!} \right) = 4x^2 - 12x^4 + \dfrac{116}{9}x^6 - \dfrac{44}{5}x^8 + \cdots$

23. $\ln(1 + \sin x) = \sum_{k=1}^{\infty} (-1)^{k+1} \dfrac{(\sin x)^k}{k}$

$= \left(x - x^3/6 + \cdots \right) - \dfrac{1}{2} \left(x^2 - x^4/3 + \cdots \right) + \dfrac{1}{3} \left(x^3 + \cdots \right) - \dfrac{1}{4} \left(x^4 + \cdots \right)$

$= x - x^2/2 + x^3/6 - x^4/12 + \cdots$

25. $\sum_{k=0}^{\infty} \dfrac{x^k}{(k + 1)!} = \dfrac{e^x - 1}{x}$

27. $\sum_{k=1}^{\infty} \dfrac{(2x)^k}{k} = -\ln(1 - 2x)$ [NOTE: $-\ln(1 - x) = \int (1 - x)^{-1}\, dx = \int \left(\sum_{k=0}^{\infty} x^k \right) dx = \sum_{k=1}^{\infty} \dfrac{x^k}{k}.$]

29. $\dfrac{\sin x}{x} = \sum_{k=0}^{\infty} (-1)^k \dfrac{x^{2k}}{(2k + 1)!} = 1 - x^2/3! + x^4/5! - x^6/7! + \cdots \implies \lim_{x \to 0} \dfrac{\sin x}{x} = 1$

31. $\dfrac{1 - \cos x}{x} = x^{-1} \left(1 - \sum_{k=0}^{\infty} (-1)^k \dfrac{x^{2k}}{(2k)!} \right) = \sum_{k=1}^{\infty} (-1)^{k+1} \dfrac{x^{2k-1}}{(2k)!} \implies \lim_{x \to 0} \dfrac{1 - \cos x}{x} = 0$

33. $\dfrac{\arctan x}{x} = x^{-1}\displaystyle\sum_{k=0}^{\infty}(-1)^k\dfrac{x^{2k+1}}{2k+1} = \sum_{k=0}^{\infty}(-1)^k\dfrac{x^{2k}}{2k+1} \implies \lim_{x\to 0}\dfrac{\arctan x}{x} = 1$

35. $\dfrac{\ln(1+x) - x}{x^2} = \displaystyle\sum_{k=0}^{\infty}(-1)^{k+1}\dfrac{x^k}{k+2} \implies \lim_{x\to 0}\dfrac{\ln(1+x) - x}{x^2} = -\dfrac{1}{2}$

37. $\displaystyle\lim_{x\to 1}\dfrac{\ln x}{x-1} = \lim_{w\to 0}\dfrac{\ln(1+w)}{w} = \lim_{w\to 0}\left(\sum_{k=0}^{\infty}(-1)^k\dfrac{w^k}{k+1}\right) = 1$

39. (a) $\dfrac{1}{1+x^4} = \displaystyle\sum_{k=0}^{\infty}(-1)^k x^{4k}$

   (b) $(-1, 1)$

   (c) $\displaystyle\int f(x)\,dx = \int\dfrac{dx}{1+x^4} = \int\left(\sum_{k=0}^{\infty}(-1)^k x^{4k}\right)dx = \sum_{k=0}^{\infty}(-1)^k\dfrac{x^{4k+1}}{4k+1}.$ Therefore, $\displaystyle\int_0^{0.5}f(x)\,dx = \sum_{k=0}^{\infty}(-1)^k\dfrac{(0.5)^{4k+1}}{4k+1}.$

   Since $\dfrac{(0.5)^9}{9} < 0.001,\; \displaystyle\int_0^{0.5}f(x)\,dx \approx \sum_{k=0}^{1}(-1)^k\dfrac{(0.5)^{4k+1}}{4k+1} = \dfrac{79}{160} \approx 0.49375.$

41. $\displaystyle\int_0^1\cos\left(x^2\right)dx = \int_0^1\left(\sum_{k=0}^{\infty}\dfrac{(-1)^k x^{4k}}{(2k)!}\right)dx = \sum_{k=0}^{\infty}\dfrac{(-1)^k x^{4k+1}}{(4k+1)\cdot(2k)!}\Big|_0^1 = \sum_{k=0}^{\infty}\dfrac{(-1)^k}{(4k+1)\cdot(2k)!} \approx \sum_{k=0}^{1}\dfrac{(-1)^k}{(4k+1)\cdot(2k)!} = \dfrac{9}{10}$
   within 0.005.

43. The power series representation of $\ln(1+x)$ is a convergent alternating series when $0 < x < 1$:
   $\ln(1+x) = \displaystyle\sum_{k=1}^{\infty}\dfrac{(-1)^{k+1}x^k}{k}.$ Because the partial sums of a convergent alternating series alternately overestimate and underestimate the limit, the first two partial sums bracket the value of $\ln(1+x)$.

45. The power series representation of $f(x) = \ln(1+x)$ converges on the interval $(-1, 1]$. The power series representation for
   $f'(x) = 1/(1+x)$ converges on the interval $(-1, 1)$. Thus, although both power series have radius of convergence 1, they
   have different intervals of convergence.

47. $y = 2e^x = \displaystyle\sum_{k=0}^{\infty}\dfrac{2x^k}{k!} \implies y(0) = 2$ and $y' = \displaystyle\sum_{k=0}^{\infty}\dfrac{2kx^{k-1}}{k!} = \sum_{k=1}^{\infty}\dfrac{2x^{k-1}}{(k-1)!} = \sum_{k=0}^{\infty}\dfrac{2x^k}{k!} = y$

49. $y = \sin x = \displaystyle\sum_{k=0}^{\infty}(-1)^k\dfrac{x^{2k+1}}{(2k+1)!}$ and $y' = \displaystyle\sum_{k=0}^{\infty}(-1)^k\dfrac{(2k+1)x^{2k}}{(2k+1)!} = \sum_{k=0}^{\infty}(-1)^k\dfrac{x^{2k}}{(2k)!}$, so
   $y'' = \displaystyle\sum_{k=0}^{\infty}(-1)^k\dfrac{(2k)x^{2k-1}}{(2k)!} = \sum_{k=1}^{\infty}(-1)^k\dfrac{x^{2k-1}}{(2k-1)!} = \sum_{k=0}^{\infty}(-1)^{k+1}\dfrac{x^{2k+1}}{(2k+1)!} = -y$

51. $\dfrac{1}{1-x} = -\dfrac{1}{1+(x-2)} = -\displaystyle\sum_{k=0}^{\infty}(-1)^k(x-2)^k \implies a_k = (-1)^{k+1}.$

53. $\displaystyle\int_0^{\infty}e^{-t}\sin(xt)\,dt = \int_0^{\infty}e^{-t}\left(\sum_{k=0}^{\infty}(-1)^k\dfrac{(xt)^{2k+1}}{(2k+1)!}\right)dt = \sum_{k=0}^{\infty}(-1)^k\dfrac{x^{2k+1}}{(2k+1)!}\left(\int_0^{\infty}e^{-t}t^{2k+1}\,dt\right) =$
   $\displaystyle\sum_{k=0}^{\infty}(-1)^k x^{2k+1} = \dfrac{x}{1+x^2}$ when $|x| < 1$

55. (a) Since $\displaystyle\lim_{n\to\infty}\left|\dfrac{a_{n+1}}{a_n}\right| = \lim_{n\to\infty}\dfrac{|r-n|\cdot|x|}{n+1} < 1$ when $|x| < 1$, the series defining f converges when $|x| < 1$.

   (b)

(c) $g'(x) = \dfrac{f'(x)}{(1+x)^r} - \dfrac{rf(x)}{(1+x)(1+x)^r} = \dfrac{f'(x)}{(1+x)^r} - \dfrac{(1+x)f'(x)}{(1+x)(1+x)^r} = 0$

(d) The result in part (c) implies that g is a constant function. Since $g(0) = 1$, $g(x) = 1$ and so $f(x) = (1+x)^r$.

57. $g(x) = \sqrt[3]{1-x^2} \approx 1 - x^2/3 - x^4/9 - 5x^6/81$

59. $g(x) = \arcsin x = \displaystyle\int \dfrac{dx}{\sqrt{1-x^2}} = \int (1-x^2))^{-1/2}\, dx = x + x^3/6 + 3x^5/40 + 5x^7/112$

11.7   Maclaurin and Taylor series

1. (a) The Maclaurin series representation of f is the polynomial expression used to define f: $x^4 - 12x^3 + 44x^2 + 2x + 1$.

(b) $f(x) = 160 + 50(x - 3) - 10(x - 3)^2 + (x - 3)^4$.

3. $K_{n+1} = e^x$ so $\left| e^x - P_n(x) \right| \leq \dfrac{e^x \cdot |x|^{n+1}}{(n+1)!} \to 0$ as $n \to \infty$.

5. (a) $\dfrac{1}{2 + x} = \dfrac{1}{2} \cdot \dfrac{1}{1 + (x/2)} = \dfrac{1}{2} \cdot \dfrac{1}{1 - (-x/2)} = \dfrac{1}{2} \sum_{k=0}^{\infty} \left(\dfrac{-x}{2} \right)^k = \sum_{k=0}^{\infty} \dfrac{(-1)^k x^k}{2^{k+1}}$

(b) The coefficient of x^{259} in the Maclaurin series representation of $f(x)$ is $f^{(259)}(0)/259!$. Thus, $f^{(259)}(0) = -259!/2^{260}$.

7. (a) $f(0) = 0$, $f'(0) = \sec^2(0) = 1$, $f''(0) = 0$, $f'''(0) = 2$, $f^{(4)}(0) = 0$, and $f^{(5)}(0) = 16$ so $\tan x = x + \dfrac{2}{3!}x^3 + \dfrac{16}{5!}x^5 + \cdots$.

(b) According to Theorem 13, the magnitude of the approximation error is no greater than $\dfrac{K_6}{6!} = \dfrac{36472}{6!} \approx 50.655$. (This error bound is much too pessimistic; the actual approximation error is $|(1 + 2/3 + 16/120) - \tan 1| \, approx 0.091$.)

9. Since $f'(0) > 0$, the coefficient of x in the Maclaurin series representation of f must be positive; the coefficient of x in the series given is negative.

11. From the information given, $f(x) = 26 + 22x - \dfrac{16}{2!}x^2 + \dfrac{12}{3!}x^3 + \cdots = 26 + 22x - 8x^2 + 2x^3 + \cdots$.

Since $P_3(1) = 42$, Theorem 13 implies that $|f(x) - 42| < \dfrac{7}{4!} = \dfrac{7}{24}$. From this it follows that

$$ (b) \; 42 - \dfrac{7}{24} = \dfrac{1001}{24} \leq f(1) \leq \dfrac{1015}{24} = 42 + \dfrac{7}{24} \; (a). $$

12.1   Differential equations: the basics

1. If $y = Ce^x$ and $x = 0$, then $y = Ce^0 = C$. This means that the graph of $y = Ce^x$ has y-intercept C. The graphs in Example 1 do exhibit this property.

3. (a) Check directly that $y' = k(y - T)$:

$$
\begin{aligned}
y'(t) &= Ake^{kt}; \\
k(y - T) &= k(T + Ae^{kt} - T) = Ake^{kt},
\end{aligned}
$$

as claimed.

(b) The situation described in the problem says, in DE language, that

$$
y'(t) = k(y(t) - 65); \qquad y(0) = 10; \qquad y(5) = 30.
$$

By (a), $y(t) = 65 + Ae^{kt}$ solves our DE. Now we'll use what we know to find values for A and k:

$$
\begin{aligned}
y(0) = 10 &\implies 10 = 65 + A \implies A = -55; \\
y(5) = 45 &\implies 45 = 65 - 55e^{5k} \implies k \approx -0.2023.
\end{aligned}
$$

Thus $y(t) = 65 - 55e^{-0.2023t}$ solves our DE. To answer the questions raised above we solve $y(t) = 60$ and $y = 64.9$ for t. The answers are (approximately) 11.85 weeks and 31.19 weeks, respectively. The following graph shows the initially promising but ultimately dispiriting progress of the campaign:

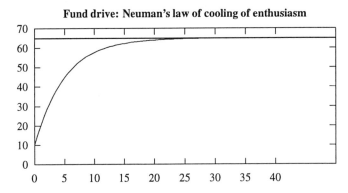

Fund drive: Neuman's law of cooling of enthusiasm

12.2   Slope fields: solving DE's graphically

1. (a) The solution curves are "parallel" to each other in the sense that they differ from each other only in their *horizontal* position. Thus, e.g., all the curves have the same slope where $y = 2$.

    (b) It *does* appear that each of the five "upper" curves has the same slope when $y = 3$. Carefully draw a tangent line to any one of the curves at the appropriate point; measure its slope. The result should be 3 (or very close to 3).

    The answer *could* have been predicted in advance. The fact that each curve is a solution to the DE $y' = y$ means precisely that when $y = 3$, $y' = 3$, too.

    (c) At the level $y = -4$, each curve has slope -4. Again, this is exactly what the DE predicts.

    (d) All curves appear to be very nearly *horizontal* near $y = 0$. The only solution curve that actually touches the line $y = 0$ is the solution curve $y = 0$ itself. Appropriately, this curve has slope 0 everywhere.

3. (a) The straight line is $y = -t - 1$; it corresponds to $C = 0$.

    (b) The curve $y = 5e^t - t - 1$ passes through $(0, 4)$. One way to tell this is to solve the equation $5 = Ce^0 - 0 - 1$ for C.

    (c) At $(0, 4)$ the curve mentioned above has slope 4. This can be found (approximately) graphically by looking at slope, or symbolically by reading the DE.

    (d) The line $y + t = 0$ (aka $y = -t$) crosses each of the "upper" four solution curves at a *stationary point*, i.e., a point where the slope is zero. This happens because, as the DE demands, $y' = 0$ whenever $y + t = 0$.

    (e) The line $y + t = -3$ (aka $y = -t - 3$) crosses the "lower" four solution curves at four points. At each of these points, the solution curve in question has the same slope: -3. This is as it should be. As the DE requires at such points, $y' = y + t = -3$.

    (f) Any line with slope -1 crosses the solution curves at points of *equal slope*. As in the previous two parts, this is because the DE requires it.

    (g) Moving clockwise from upper left, the curves correspond to the C-values $500, 50, 5, 0.2, 0, -0.2, -5, -50, -500$.

5. In the slope field for $y' = y$, all ticks at the same *vertical position* are parallel. (That's because all such ticks have the same y-value.) In the slope field for $y' = 2t - 5$, all ticks at the same *horizontal position* are parallel. (That's because all such ticks have the same t-value.)

12.3   Euler's method: solving DE's numerically

1. (a) A table helps keep track of results:

step	t	y'	y
0	0.00	0	0
1	0.25	0.2474	0
2	0.50	0.4794	0.06185
3	0.75	0.6816	0.1818
4	1.00	0.8415	0.3522

(b) The left rule with 4 subdivisions, applied to $I = \int_0^1 \sin(t)\,dt$, gives

$$L_4 = \frac{\sin(1/4)}{4} + \frac{\sin(1/2)}{4} + \frac{\sin(3/4)}{4} \approx 0.3520.$$

(c) The function $y(t) = 1 - \cos t$ solves the DE exactly. Thus, exactly, $y(1) = 1 - \cos 1 \approx 0.4597$.

(d) $\displaystyle\int_0^1 \sin(t)\,dt = -\cos t]_0^1 = -\cos 1 + \cos 0 = 1 - \cos 1 \approx\approx 0.4597$. The error committed by L_4, therefore, is

$$|I - L_4| \approx 0.4597 - 0.3520 = 0.1077.$$

(e) Here $y(t) = 1 - \cos t$, and $Y(t)$ is the function tabulated above. Thus we get:

t	0.00	0.25	0.50	0.75	1.00
$Y(t)$	0	0	0.0618	0.1818	0.3522
$y(t)$	0	0.0311	0.1224	0.2683	0.4597

(f) To plot $y(t)$ and $Y(t)$ on one pair of axes, we use the formula $1 - \cos t$ for y, and "connect the dots" for Y:

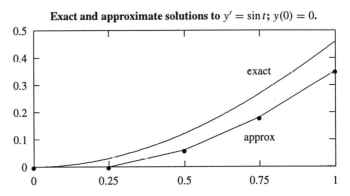

Exact and approximate solutions to $y' = \sin t$; $y(0) = 0$.

Notice that the two functions start out together, but spread apart as t increases.

3. (a) It's not hard to guess that the function $y(t) = 3t$ solves the IVP above. Thus $y(1) = 3$, $y(2) = 6$, $y(3) = 9$, $y(4) = 1$ $y(5) = 15$.

(b) In this case, Euler's method gives *exact* values; the Euler estimates commit *no* error.

(c) Euler's method pretends, in effect, that y' remains constant over small intervals. In this case, y' *is* constant, so Euler method commits no error.

Euler's method will behave this way (i.e., commit *no* error) whenever y' is a constant function.

5. One does the "easy calculation" by checking explicitly that both the DE and the initial condition are satisfied. Since $y(t) = 70 + 120e^{-0.1t}$, it's easy to check that

$$y'(t) = -12e^{-0.1t} = -0.1 \cdot 120 \cdot e^{-0.1t}) = -0.1 \cdot (y - 70).$$

Thus, the DE does hold as advertised. Also, $y(0) = 70 + 120 = 190$, so the initial condition holds, too. Finally,

$$y(5) = 70 + 120e^{-0.1 \cdot 5} = 70 + 120e^{-0.5} \approx 142.78.$$

7. (a) Work quickly shows that given *initial* population 0, the population *remains* at 0. Nothing happens.

(b) In biological terms, the situation is simply that without any initial breeding members a population can't grow. No parents; no children.

12.4 Separating variables: solving DE's symbolically

1. (a) That all curves "start" at (0, 190) means that all cups of coffee are at 190 degrees at time 0.

    (b) In each case, T_r describes the (stable) *room* temperature. Thus the nine coffee cups are in rooms of different temperatures.

    (c) One way to find the values of T_r that correspond to C_1, C_4, and C_9 is just to observe that T_r represents the "long-run" temperature—the temperature to which the coffee tends over a long period of time. Looking at the right-hand parts of the graphs lets us read off these numbers for C_1 and C_4. Thus C_1 corresponds to $T_r = 100$; C_4 corresponds to $T_r = 70$. Not enough of C_9 appears for this to work very well. Instead, we can use the fact that for C_9, $y(10) = 70$. From this it follows:
    $$y(10) = (190 - T_r)e^{-1} + T_r = 70 \implies T_r \approx 0.$$

    (d) C_9 is cooling fastest at $t = 10$; of all the curves, C_9 has steepest downward slope at $t = 10$.

    (e) Curve C_4 solves the IVP $y' = -0.1(y - 70)$; $y(0) = 190$. In the same spirit, C_1 solves the IVP $y' = -0.1(y - 100)$; $y(0) = 190$, and C_9 solves the IVP $y' = -0.1(y - 0)$; $y(0) = 190$. (The DE's differ only in the values of T_r.)

    (f) Estimating the *slope* of the curve C_1 at the point $t = 10$ gives $y'(10) \approx -3.5$ degrees per minute.

    (g) The DE for C_1 is (as we found above) $y' = -0.1(y - 100)$. The graph C_1 shows that when $t = 10$, $y \approx 135$, so, according to the DE, $y' = -0.1(135 - 100) = -3.5$ degrees per minute. (This agrees with the previous part, of course.)

    (h) Estimating the *slope* of the curve C_9 at the point $t = 10$ gives $y'(10) \approx -7$ degrees per minute.

    (i) The DE for C_9 is (as we found above) $y' = -0.1y$. The graph C_9 shows that when $t = 10$, $y \approx 70$, so, according to the DE, $y' = -0.1 \cdot 70 = -7$ degrees per minute. (This agrees with the previous part, as it should.)

    (j) Looking carefully at the picture. A curve that represents coffee that starts at 190 degrees and is at 100 degrees after 20 minutes would fall about *halfway* between C_2 and C_3. In particular, C_2 and C_3 correspond to room temperatures of 90 and 80 degrees, respectively. Thus the desired curve would correspond to room temperature of about 85 degrees.

3. (a) If $P < C$, then $P' > 0$. This means that the population *increases*—as it should, since carrying capacity hasn't been reached.

    (b) If $P > C$, then $P' < 0$. This means that the population *decreases*—as it should, since carrying capacity has been exceeded.

    (c) If $P = C$, then $P' = 0$. This means that the population is *stable*—as it should be, when exactly at capacity.

5. Solving algebraically for k gives $k = -\dfrac{\ln(\frac{19}{49})}{1000}$; an approximate decimal equivalent is 0.000947, as claimed.

7. (a) The exact values of k for curves P_1 to P_4 are, respectively, 2, 1.5, 1, and 0.5. (Thus the values of $-k$ are, respectively, -2, -1.5, -1, and -0.5.)

    The k-values can be found—approximately—by reading the graphs. The graph of P_1, for example, seems to pass through the point (2, 750). This gives an equation we can solve for k, as follows:
    $$P_2(2) = \frac{1000}{19e^{-2k} + 1} \approx 750 \implies k = \frac{\ln 57}{2} \approx 2.02.$$

    Values of k for the other curves are found similarly.

    (b) P_1 corresponds to the "hottest" rumor, P_4 to the "coolest." The graph shapes agree—hotter rumors spread faster.

9. Separating variables in the DE gives
    $$\frac{dP}{dt} = -0.00000556 P(P - 10000) \implies \int \frac{dP}{P(P - 1000)} = -\int 0.00000556 dt.$$

    Integrating both sides gives
    $$\frac{\ln|P|}{10000} + \frac{\ln|P - 10000|}{10000} = -0.00000556t + C,$$

for some constant C. Setting $P = 1000$ and $t = 0$ gives

$$C = -\frac{\ln 1000}{10000} + \frac{\ln 9000}{10000} = -\frac{\ln 9}{10000} \approx -0.0002197.$$

Setting $t = 10$ and solving for P gives $P(10) \approx 1622$—not much different from what we've seen in other sections.

13.1 Polar coordinates and polar curves

1. (a) $(\sqrt{2}, \pi/4)$, $(\sqrt{2}, -7\pi/4)$, $(-\sqrt{2}, -3\pi/4)$, $(-sqrt2, 5\pi/4)$
   (b) $(\sqrt{2}, 3\pi/4)$, $(\sqrt{2}, -5\pi/4)$, $(-\sqrt{2}, 7\pi/4)$, $(-\sqrt{2}, -\pi/4)$
   (c) $(2, \pi/3)$, $(2, -5\pi/3)$, $(-2, 4\pi/3)$, $(-2, -2\pi/3)$
   (d) $(2, 2\pi/3)$, $(2, -4\pi/3)$, $(-2, -\pi/3)$ $(-2, 5\pi/3)$
   (e) $(\pi, 0)$, $(-\pi, \pi)$, $(-\pi, \pi)$, $(\pi, 2\pi)$
   (f) $(\pi, \pi/2)$, $(\pi, -3\pi/2)$, $(-\pi, -\pi/2)$, $(-\pi, 3\pi/2)$

3. (a) $(\sqrt{2}, \sqrt{2})$
   (b) $(\sqrt{2}, \sqrt{2})$
   (c) $(\sqrt{3}/2, 1/2)$
   (d) $(42, 0)$
   (e) $(a, 0)$
   (f) $(-a, 0)$

5. Points of the form $(a, 0)$ where $a \in \mathbb{R}$ have this property.

7. (a)

θ	0	$\frac{\pi}{6}$	$\frac{\pi}{3}$	$\frac{\pi}{2}$	$\frac{2\pi}{3}$	$\frac{5\pi}{6}$	π	$\frac{7\pi}{6}$	$\frac{4\pi}{3}$	$\frac{3\pi}{2}$	$\frac{5\pi}{3}$	$\frac{11\pi}{6}$	2π
r	2	1.866	1.5	1	0.5	0.134	0	0.134	0.5	1	1.5	1.866	2

   (c) The cardioid is symmetric with respect to the x-axis.

9.

11. (a) The polar points $(1, 0)$, $(1, 2\pi)$, and $(-1, \pi)$ all represent the Cartesian point $(1, 0)$. (Use $x = r\cos\theta$ and $y = r\sin\theta$.)
    (b) The polar points $(-1, \pi/4)$, $(-1, 9\pi/4)$, and $(1, 5\pi/4)$ all represent that Cartesian point $(-\sqrt{2}/2, -\sqrt{2}/2)$.
    (c) $(\sqrt{2}, \pi/4 + 2k\pi)$ and $(-\sqrt{2}, \pi/4 + (2k-1)\pi)$

13. (a) $r = 3$
    (b) $r = 4\csc\theta$
    (c) $\tan\theta = 2$
    (d) $r = 2\cos\theta$

15.

3.2   Calculus in polar coordinates

1.

3. (b) $\dfrac{dy}{dx} = \dfrac{\sin\theta + \theta\cos\theta)}{\cos\theta - \theta\sin\theta}$. Thus, the spiral has a horizontal tangent line wherever $\theta = -\tan\theta$ and a vertical tangent line wherever $\theta = \cot\theta$.

(c) The polar point $(1, 1)$ is the point $(\cos 1, \sin 1)$ in Cartesian coordinates. The slope of the tangent line at the polar point $(1, 1)$ is $m = (\sin 1 + \cos 1)/(\cos 1 - \sin 1) \approx -4.588$. Thus, the equation of the desired tangent line is $y = m(x - \cos 1) + \sin 1 \approx -4.59(x - 0.54) + 0.84$.

5. (b) $\dfrac{dy}{dx} = \dfrac{-a\sin^2\theta + \cos\theta + a\cos^2\theta}{-2a\sin\theta - \sin\theta}$. The limaçon has a vertical tangent whenever the denominator in this expression is zero (i.e., when $\theta = 0$ or $\cos\theta = -1/2a$). Thus, there will be three vertical tangent lines if and only if $|a| \le 1/2$.

7. (a) area $= \dfrac{1}{2}\displaystyle\int_0^{\pi/6} f(\theta)^2\, d\theta + \dfrac{1}{2}\int_{11\pi/6}^{2\pi} f(\theta)^2\, d\theta = \dfrac{1}{2}\int_{-\pi/6}^{7\pi/6} f(\theta)^2\, d\theta = 2\pi + 3\sqrt{3}/2$

(b) area $= \dfrac{1}{2}\displaystyle\int_{7\pi/6}^{11\pi/6} f(\theta)^2\, d\theta = \pi - 3\sqrt{3}/2$

(c) area $= 1$

(d) area $= m$

9. (a) One leaf lies between $\theta = -\pi/2n$ and $\theta = \pi/2n$. The area of this leaf is $\dfrac{1}{2}\displaystyle\int_{-\pi/2n}^{\pi/2n} \cos^2(n\theta)\, d\theta = \pi/4n$.

(b) The area of all n leaves is $\pi/4$ (i.e., one-fourth of the area of the circle $r = 1$).

11. area $= \left(e^{4\pi} - 1\right)/4$

13. area $= \dfrac{1}{2}\displaystyle\int_{-\pi/3}^{\pi/3} d\theta - \dfrac{1}{2}\int_{-\pi/3}^{\pi/3}\left(\tfrac{1}{2}\sec\theta\right)^2 d\theta = \dfrac{\pi}{3} - \dfrac{\sqrt{3}}{4}$

15.